Food Structure and Behaviour

FOOD SCIENCE AND TECHNOLOGY

A SERIES OF MONOGRAPHS

A complete list of the books in this series is available from the publishers on request

Food Structure and Behaviour

Edited by

J. M. V. BLANSHARD
Department of Applied Biochemistry and Food Science, Nottingham University, UK

and

P. LILLFORD
Unilever Research, Bedford, UK

1987

ACADEMIC PRESS
Harcourt Brace Jovanovich, Publishers
London Orlando San Diego New York
Austin Boston Sydney Tokyo Toronto

ACADEMIC PRESS LIMITED
24–28 Oval Road, London NW1 7DX

United States Edition published by
ACADEMIC PRESS, INC.
Orlando, Florida 32887

British Library Cataloguing in Publication Data
Food structure and behaviour.—(Food science and technology).
1. Food 2. Rheology 3. Food texture
I. Blanshard, J.M.V. II. Lillford, P.
III. Series
664 TP368

ISBN 0-12-104230-8

Filmset by Eta Services (Typesetters) Ltd, Beccles, Suffolk
Printed in Great Britain at the University Press, Cambridge

Contributors

A. G. Atkins, Department of Engineering, University of Reading, Whiteknights, Reading RG6 2AY, UK

J. M. V. Blanshard, Department of Applied Biochemistry and Food Science, Faculty of Agricultural Science, School of Agriculture, University of Nottingham, Sutton Bonington, Loughborough LE12 5RD, UK

D. Burfoot, AFRC Meat Research Institute, Langford, Bristol BS18 7DY, UK

A. H. Clark, Unilever Research, Colworth Laboratory, Colworth House, Sharnbrook, Bedford MK44 1LQ, UK

D. F. Darling, Unilever Research, Colworth Laboratory, Colworth House, Sharnbrook, Bedford MK44 1LQ, UK

S. F. Edwards, Cavendish Laboratory, University of Cambridge, Cambridge CB3 0HE, UK

F. Franks, PAFRA Ltd, Biopreservation Division, 150 Cambridge Science Park, Cambridge CB4 4GG, UK

J. Garside, Department of Chemical Engineering, University of Manchester Institute of Science and Technology, PO Box 88, Manchester M60 1QD, UK

P. B. Johns, Department of Electrical and Electronic Engineering, University of Nottingham, University Park, Nottingham NG7 2RD, UK

P. J. Lillford, Unilever Research, Colworth Laboratory, Colworth House, Sharnbrook, Bedford MK44 1LQ, UK

P. B. McNulty, University College Dublin, Department of Agricultural Engineering, Earlsfort Terrace, Dublin 2, Ireland

S. H. Pulko, Department of Electronic Engineering, University of Hull, Hull HU6 7RX, UK

P. P. Purslow, AFRC Institute of Food Research, Bristol Laboratory, Langford, Bristol BS18 7DY, UK

P. Richmond, AFRC Institute of Food Research, Colney Lane, Norwich NR4 7UA, Norfolk

A. C. Smith, AFRC Institute of Food Research, Norwich Laboratory, Colney Lane, Norwich NR4 7UA, UK

P. Walstra, Dept of Food Science, Wageningen Agricultural University, De Dreijen 12, 6703 BC Wageningen, The Netherlands

Preface

The scientific study of food has undoubtedly gathered momentum over the past 10–20 years and gained a well-earned academic respectability. A number of factors have contributed to these developments. On the one hand there is the intrinsic interest in investigating materials which are of immediate significance to daily life—materials which by any standard pose a substantial challenge by reason of their scientific complexity. Also there are the commercial pressures at local, national, and international levels, which promote research. In some countries at least, the slimming down of corporate research budgets and the intensifying of international competition has encouraged more government financed research in food which again has raised the profile and importance of this area.

However, for the scientist, the enthralling nature of food research lies not only in the immediacy of its application but also in the challenge of comprehending, analyzing and modelling such systems and relating the molecular composition and configuration to the perceived macroscopic structure. A further feature which, though obvious on reflection but not immediately apparent, is that foodstuffs are rarely in an equilibrium state. Processing and preservation techniques therefore need to recognize this characteristic and handle the associated problems and possibilities accordingly.

The contents of this book, which were the substance of a Post-Experience Course organized under the auspices of the Royal Society of Chemistry and held at the Nottingham University Faculty of Agricultural Science at Sutton Bonington, are an attempt to grapple with some of these challenging issues. It will be evident that the treatment could not be fully comprehensive, but the intention has been to explore certain critical areas which are of importance to the food industry and show considerable potential in terms of scientific understanding and development. Since the chapters were constructed with a didactic purpose in view, any one may be read in isolation but the overall aim is that the reader should gain a fuller insight into the concepts of the physical and physicochemical framework which is gradually being constructed and which undoubtedly will assist the food processing industry.

Those who attended had the benefit of detailed discussions and tutorials in some of these areas. They are not included in the book but the bibliography associated with each chapter will permit the perceptive reader to pursue each subject to a greater depth.

The editors would like to express their warm appreciation to the authors for their painstaking contributions and to Dr. A. D. Ashmore and his staff of the Royal Society of Chemistry for their administrative expertise before and during the course.

J. M. V. Blanshard
P. J. Lillford

Contents

1 Gels and Networks in Practice and Theory

S. F. EDWARDS, P. J. LILLFORD and J. M. V. BLANSHARD

University of Cambridge, Unilever Research, Bedford, and University of Nottingham

1 INTRODUCTION

Gels are found in foods and provide easily recognized, though widely varying textures; some are highly acceptable (steak), some specialist (jelly) and some are troublesome (gristle).

Recently, the role of ingredients in building food structures has been recognized, so that gelation has been defined as a "functional" property of the ingredients. As little detailed knowledge was available about the molecular control of the gelation process, emphasis in the 60s and 70s was placed on understanding ingredient behaviour, with enormous success in some areas and considerable frustration in others. At the same time, an alternative approach, starting with known finished foods, examining the contribution

FOOD STRUCTURE AND BEHAVIOUR
ISBN 0-12-104230-8

and function of the ingredients within their structure has been followed. In some product areas these approaches met with a successful outcome; in other situations they bypass each other because there is an inadequate recognition and matching of the distinctive contributions of components to the complex physical texture of the whole product. In some other cases, these approaches are still struggling to reach each other. This chapter will try to exemplify work from both ends.

2 GELATION MECHANISMS OF FOOD COMPONENTS

Because food manufacture is a large-scale business, the ingredients must be cheap and in copious supply. Gelling agents can be categorized as "cold-setting" and "heat-setting", polymers. If we look at their natural origin it is not surprising that the cold-setting polymers provide structures in biology. To do so they must form networks under physiological conditions which by food processing standards are "cold". Heat-setting polymers derive from a variety of sources, their gelation is largely coincidental to their biological function and is an interesting high-temperature artefact.

This at least provides a convenient classification to discuss the known gelation mechanisms of food polymers.

A Structural polymers

First, there are some structural polymers which the food industry has found difficult to manipulate. They provide the mechanical stability of land plants and animals and are either indigestible (cellulose) or unbreakable in the mouth (elastin, keratin, and some collagens). The three protein networks have one common element, that of naturally occurring, permanent, covalent cross-links. The cross-linking makes the networks undigestible.

Useful network forming properties have been obtained from these materials by chemical modification.

1 CELLULOSE DERIVATIVES

Chemical substitution by hydrophobic groups to form methyl- or hydroxy-propyl cellulose produces cold water solubility. Elevation of the temperature causes thickening and the formation of mechanically weak gels, presumably, by the formation of hydrophobic interchain contacts. Carboxymethylation also induces water solubility. At high polymer concentration and low degrees of substitution weak thixotropic gels form. Presumably, the interchain contacts are formed by desolvation of glucose residues.

2 GELATIN

The extent of covalent cross-linked-linking of collagen depends upon the muscle source and animal age. For example, fish collagen is largely hot-water soluble, as is a major fraction of mammalian collagen from young animals. This soluble material gels on cooling by the restoration of triple helical junction zones similar to those of natural tropocollagen. The majority of food gelatin is derived from much cruder sources, namely by partial alkaline or acid hydrolysis of animal hides. The resultant soluble hydrolysates are extremely heterogeneous in molecular weight, chain-length, and isoelectric point; (acid gelatin having an p$I \sim 9$, and limed gelatin p$I \sim 5$). Because of the complexity of the molecular structures, commercial gelatins are characterised in terms of their gels strength (Bloom value). In fact, the cold gelatin process of these complex mixtures probably never reaches equilibrium, a slow annealing process taking place after the formation of the initial gel.

3 AGAR AND CARRAGEENANS

These polymers are derived from red seaweeds and have similar structural units (Fig. 1). They differ primarily in their degree of sulphation and hence their surface charge. The molecular conformational changes accompanying gelation are under intensive study.

Agarose on cooling, forms double helical structures which subsequently or concurrently form networks via aggregation. It is generally agreed that the mechanism of gelation is in principle the same for carrageenans. Their

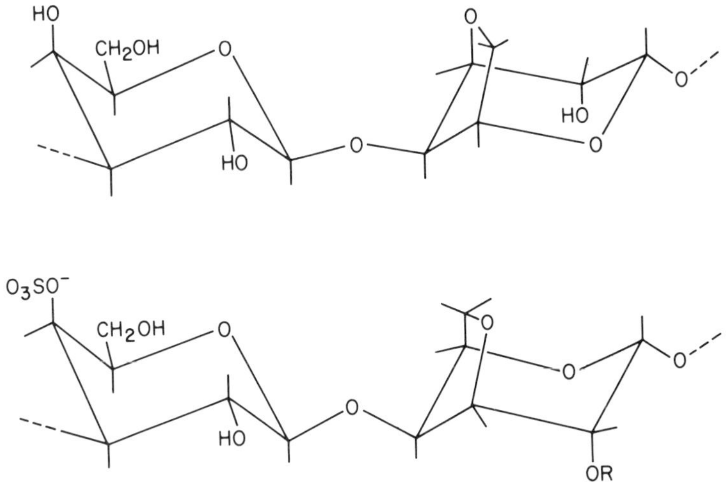

Fig. 1

polyelectrolyte nature, however, means that counterion type and concentrations have a major effect on both double helix formation and subsequent aggregation. In general, the degree of sulphate substitution can be related to the extent of aggregation, hence the turbidity, brittleness and resistance to freeze/thaw damage decreases as the sulphate substitution increases. Commercial samples frequently exploit this property by blending carrageenan types for specific gel properties. Problems are encountered in predicting the behaviour of these materials in food formulation because of their counterion sensitivity. This can influence gelling properties and derives from competition for ions with other polyelectrolytes (e.g. milk proteins).

4 ALGINATES AND PECTINS

Alginates are linear polymers of mannuronic (M) and guluronic (G) acids found in brown seaweeds. The two acids are present as blocks of like acids and alternating sequences. The polymers will not gel in the absence of divalent cations and the interchain association is associated with a cooperative association of G blocks of greater than 20 U. Hence the gelling characteristics can to some extent be predicted by examination of the M/G ratio in the initial seaweed source, and the proportion of M and G blocks. By virtue of their ionic junction zones, the alginates form thermally irreversible gels. Prediction of their properties in use is again complicated by their cation sensitivity, and great care must be taken with calcium levels of other ingredients.

Pectins are branched polysaccharides found in land plants. Essentially their backbone has a structural similarity with guluronate. The extent of methylation and the extent and types of side chain substitution vary with the source. Low methoxy and amide pectins gel in the presence of divalent cations, and the latter exhibit a degree of thermoreversibility. The cooperative unit for association for polygalacturonate appears to be 14 residues. High methoxy pectins are the mainstay of the jam industry, where gelation accompanies the addition of high levels of sucrose. Gelation can also be accomplished by the addition of glycerol and ammonium sulphate and presumably involves network formation by "salting out" of the chains by colloidal mechanism comparable to the formation of protein precipitate gels discussed later.

B Thermogelling polymers

1 STARCH

Starches are the most important reserve carbohydrate in the plant world and widely used as a structuring agent in low and high technology food processes.

As an ingredient, it is extracted from only a few species such as maize, rice, tapioca, and sago; recently air classified fractions from leguminous seeds have been produced. Starches are stored in plants in the form of 2–100 μ granules, varying in ratios of amylose (a linear glucan) and amylopectin (a branched glucan of mol wt $\sim 10^8$ daltons). The granules are unaffected by water until a "gelatinization" temperature is reached. This corresponds to crystallite melting and is subsequently followed by solvation of the macromolecules which eventually form a fibrous elastic mass by 110 °C.

In addition, amylose is insoluble at temperatures less than $\sim$150 °C. Precipitates and gels form on cooling, and gelation is favoured by accelerated cooling. Gelation also shows a marked dependence on chain length. Sols are stable at chain lengths less than 40 residues; between 40 and 80 residues, aggregates and precipitates form; above 80 residues gelation is favoured. Presumably this is a demonstration of kinetic control such that at high molecular weight the rates of chain matching are slow compared to chain/chain association. In all cases gels subsequently anneal and contract.

The gels formed from native starch are believed to involve both amylopectin and amylose but whether there is molecular interaction or phase separation is not clear. The gels are enormously shear sensitive, presumably due to easy cleavage of the amylopectin molecule. In food manufacture, all the structural and molecular states of starch are used. Even within a single product, swollen granules, partially leached granules and homogeneous structures are observed which are dispersions rather than simple networks, so that starch is usually regarded as a thickener rather than a gelling agent.

2 GLOBULAR PROTEINS

Most globular, native proteins, whether albumins or globulins can be gelled on heating, provided their initial solution concentration is sufficiently high. Unlike the structural polymers, however, the junction zones cannot be defined with respect to extended ordered structures so that gelation and gel properties are described in terms of more general principles of colloidal attraction and repulsion. At least three events can be identified in the thermal gelation process.

(i) denaturation → (ii) aggregation → (iii) crosslinking

The native state of proteins is maintained by a delicate balance of chain interaction energies involving electrostatic, hydrogen-bond and hydrophobic interactions, all of which have marked temperature derivatives but of opposing sign. At the denaturation temperature, alternative minimised energy conformations become available which for single subunit proteins involves chain reorganization and for multisubunit proteins may also involve subunit separation. In neither case, do extended "random coil" structures form. If hydrophobic association outweighs electrostatic repulsion, the second stage of interchain aggregation occurs, causing precipitation at low concentrations or gelation at higher concentrations. Lastly, slow rearrangement or annealing on cooling occurs over a timescale of days, which involves disulphide exchange, hydrogen bond "shuffling" etc. It is very important to recognize, that since the heat denatured conformation is dependent on the environment (pH, ionic strength Aw etc.), the junction points, network, and subsequent cooled structure are a function of the total history of the gelation process. Hence the order of events is critical in determining the final properties of the matrix. For example, the rheological properties of an albumin gel made by heat-setting at its isoelectric point and adjustment to a high pH, are not the same as those derived from the same protein heat set directly at the higher pH.

Another factor influencing the properties of heat set matrices is the presence of insoluble material. Food grade gelling proteins (e.g. egg albumin) are stored in bulk either frozen or dried. These processes induce insolubilization, so that the redissolved material always contains large hydrated and dispersed aggregates. Heat set gels from these materials are weaker than their counterparts from totally soluble protein.

Even if concentration is corrected for the presence of insoluble material, the reduced gel strength is not explained. Presumably insoluble material acts as nuclei for aggregation processes and predetermines the structure of the growing network.

C Networks from native protein

It is not necessary to denature proteins to form networks, many traditional food processes rely on their ability to self associate in the native state. For example, the reduction of pH by fermentive bacteria to around the isoelectric point of casein is a fundamental process in cheese manufacture. Similarly, the swelling and partial dissolution of actomyosin by salt, followed by re-aggregation at pH 4.5–5.5 is the basis of dry sausage (salami) manufacture. Similar gels can be fomed by non-fermentive means, by direct acidification or the use of glucono-δ-lactone decomposition. (The textures and flavour, however, are rarely as satisfying.) Calcium induced aggregation of soy protein is the basis of Japanese tofu manufacture. The heating process preceding precipitation is required to activate antinutritional factors rather than create the network structures.

D Networks from denatured protein

The desire to create analogue food structures from unusual protein sources has led to a new technology of protein structuring over the past 50 years. As usual, technology has advanced faster than its science so that developments are publicized initially in the patent literature and reviews of the underlying science have followed. Typically, the processes for fibre production (i.e. manufacture of cylindrical gels) involve dissolution of the native protein in a denaturant, followed by injection into a coagulating bath which removes the denaturant and promotes protein aggregation. Urea, and detergent denaturants have been used but the most successful Boyer process involves alkali denaturation followed by coagulation at high ionic strength in acidic conditions. The initial network is probably formed from essentially globular units since even at the high pH conditions of the initial spinning dope, the peptide chains are far from "random coils". Subsequent "drawing" of the fibres increases their mechanical stability and a degree of crystallinity has been detected.

Thermoplastic extrusion of proteins is also widely used to produce solid structures. The process is probably more appropriately described as network conversion rather than network formation, since the solids content of the feed material is so high that the proteins are initially in a three dimensional structure. Both native and denatured protein can be extruded but virtually no knowledge of the accompanying molecular events is available.

E Gluten

The ability of wheat proteins to form cohesive elastic networks is unique in food polymer systems. Despite the fact that cereal proteins account for the major source of protein nutrition for humans, and that most of the baking process depends on its properties, gluten remains an enigma. This is certainly due to its intractable nature with respect to molecular investigation and characterization.

The glutenin fraction of wheat responsible for the network formation comprises peptide chains which vary with wheat variety. The amino acid composition indicates that no extended, ordered, secondary structures can be formed in their native or hydrated states and this is borne out by the fact that high resolution nmr spectra can be observed from fully developed gluten whose molecular weight is effectively infinite. The only other protein system exhibiting this phenomenon of high segment mobility at high molecular weight is elastin so that these two networks are rubber-like in their structure and properties. The cross-links are generally agreed to involve disulphide linkages, and the position of the sulphydryl groups within the peptide chains appears to be critical. For example, it has been shown that one particular peptide chain is present in all bread-making "strong" wheats. Further elucidation of the molecular mechanism of the network development during work input and its subsequent change on heating requires more detailed investigation.

3 MOLECULAR NETWORKS

Now that we have identified and attempted to classify the types of molecules which form networks in food systems there still remain a number of important questions. For example, how can we physically model such complex polymers and the networks they form? Further, how can we mathematically formulate such systems so as to give useful information on the effects of modifying physical variables such as temperature, concentration, improved stress etc.? Undoubtedly the simplest models are derived from simple random polymers permanently cross-linked at random positions.

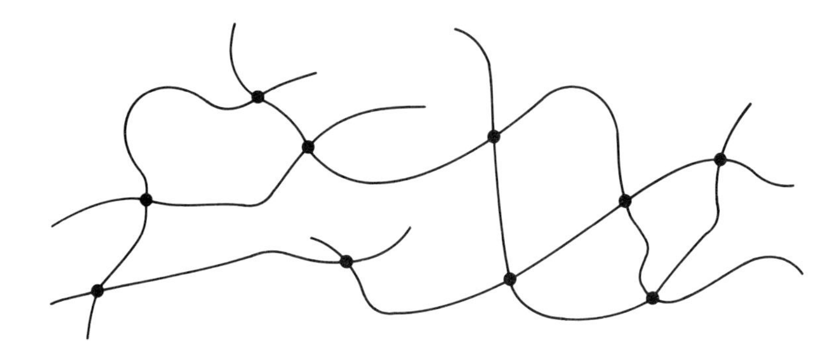

A Polymer statistics for the simplest model

Flory calculated the elastic modulus of such a system, assuming that the change in free energy associated with a deformation is due solely to entropic changes in the chains.

i.e. $$\Delta A = -TAS$$

Suppose that between two cross-links the polymer has a random flight step length b, and N steps.

When N is large, the number of configurations is

$$Z = B^N \exp\left[-\frac{3}{2b^2}\frac{R^2}{N}\right]\left(\frac{3}{2b^2 N\pi}\right)^{3/2} \tag{1}$$

where B^N is the number of configurations if the ends are free.

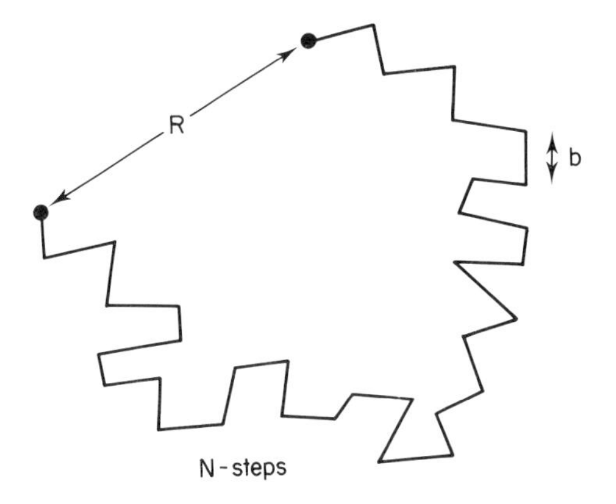

Then $S = k \log Z$.

$$S = Nk \log B + \frac{3k}{2} \log\left(\frac{3}{2b^2 N\pi}\right) \frac{3k}{2b^2} \frac{R^2}{N}$$

Now deform the system so that $X \to \lambda_1\, x$ etc.

$$Y \to \lambda_2\, y$$
$$Z \to \lambda_3\, z$$

Expression (1) becomes

$$\left[B^N \exp\left(\frac{-3}{2b^2} \frac{(\lambda_1^2 X^2 + \lambda_2^2 Y^2 + \lambda_3^2 Z^2)}{N}\right)\right] - \left(\frac{3}{2b^2 N\pi}\right)^{3/2}$$

$$\Delta S = -\frac{3k}{2b^2 N}\left[\begin{matrix}(\lambda_1^2 - 1)X^2 + (\lambda_2^2 - 1)Y^2 \\ + (\lambda_3^2 - 1)Z^2)\end{matrix}\right]$$

But what are X, Y, Z? Suppose the cross-linking took place instantly so that R is the typical distance along a random chain. The probability of a typical distance R is then,

$$\mathrm{P}(R) = \frac{Z(R)}{\int Z(R)\, \mathrm{d}^3 R} = \left(\frac{3}{2b^2 N\pi}\right)\Bigg]\left(\frac{3}{2b^2 N\pi}\right)^{3/2}$$

So the mean value of ΔS will be

$$\Delta\bar{S} = \int \Delta S(R)\mathrm{P}(R)\, \mathrm{d}^3 R$$

For simple elongation

$$\Delta\bar{S} = \frac{k}{2}(\sum (\lambda_i^2 - 1)) \text{ per segment.}$$

Hence total

$$\Delta A = \frac{kT}{2}\left(\lambda^2 + \frac{2}{\lambda} - 3\right)N$$

where N = number of segments per unit volume.

Note the force

$$F = \frac{\delta \Delta A}{\delta \lambda} = NkT\left(\lambda - \frac{1}{\lambda^2}\right)$$

is proportional to temperature, number of links and $\lambda - 1/\lambda^2$.

Experimentally all these results have some validity but are not perfect.

This simple theory has been successfully applied to the structural proteins resilin, abductin and elastin. The latter material provides the inedible, rubbery gristle of ligament and fascia in animal tissue. Even here, however, the model has required *modification* to account for polymer/solvent interactions.

B Difficulties with the simple theory

These are both physical (in other words, "how good is the model?") and mathematical ("how well did you solve it?").

1 PHYSICAL PROBLEMS

The free energy is not all entropic and internal energy terms often give contributions between 0–10%.

The cross-links are not formed instantaneously but usually take time. Once a link is in place there is an added probability that the next link will be near the first, so links tend to form clumps and are less effective.

The chains are entangled, so that, even without cross-links, melts show rubbery properties on time scales shorter than creep times.

2 MATHEMATICAL PROBLEMS

Given some decison on where the cross-links went, there is still no reason why they should move affinely, and indeed they do not. Mathematically the affine hypothesis is only valid in the limit of a large number of chains emanating from each cross-link. For the case above, the relaxation of the affine hypothesis halves the free energy (proved by much calculation).

It is not clear that the network is stable. Physically, of course, it obviously is but the model of gaussian chains without entanglements is rather like having the polymer molecules behave like stream lines in a turbulent fluid, which (without viscosity etc.) can be stirred into infinite complexity. This is a problem one would never suspect at first but it appears in any better calculation. Entanglements resolve it.

C Entanglements

To describe these, imagine all chains but one at right angles to the diagram.

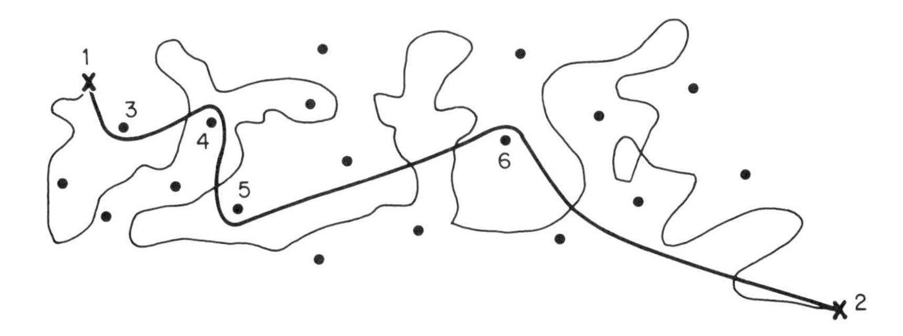

Imagine a hypothetical individual hauling in the chain at point 2, keeping it fixed at point 1. A much shorter locus results which is called the *primitive path*. It is the topological skeleton of the real polymer. One can then imagine the real polymer very roughly confined to pass through hoops at 3, 4, 5, 6, so that, again roughly speaking, 1–2 is not one segment but five (i.e. N_x in the simple formula ought to be $N_x + N_e$) where N_e is the number of entanglements and depends on the topology.

If one studies the modulus at a sufficiently high frequency, so that the polymer does not have time to slide through the entanglements, a higher modulus should be found. It is, and serves to define the effective number of entanglements. Indeed if there are no cross-links at all, the material will creep at constant stress, but at high frequency will give the number of entanglements. Thus we find the number of entanglements from the high frequency moduli; some of these are trapped by permanent cross-links.

If the number is η^{-1}, a calculation gives trapped entanglement segments contributing

$$\frac{1}{2}\sum_i \frac{\lambda_i^2(1+\eta)}{1+\eta\lambda_i^2} + \sum_i \frac{1}{2}\log(1+\eta\lambda_i^2)$$

It will be seen that this formula weakens at increasing λ; but at large enough λ, segments of chain get fully extended and a big increase in modulus is to be expected and is found.

D Temporary networks

In foods most cross-links are not covalent. Instead in hydrogen bonded materials, or those having cross-links depending on charges along chains, links can open and close by thermal agitation. Such effects happen at high frequency compared to normal stresses and strains, so a gelatin gel (for example a table jelly) creeps very slowly. It is possible to set up a theory for such repeated breaking and reforming of links. When a link breaks, entropy is gained as the free chain relaxes; when the link is reformed it notices strain changes from that point onwards. However the results of theory are very peculiar, for the loss of stress decreases with time.

E Visco elasticity

The primitive path picture suggests that the polymer sits in a statistically defined pipe, but without permanent cross-linked links can wriggle out of the old pipe, creating a new pipe. Under a sudden deformation this happens.

- **a** The pipe deforms affinely, taking the polymer with it.
- **b** The polymer shrinks down the pipe to its natural length.
- **c** The distribution of steps of the primitive path is not correct, (long ones along axis of strain, short ones perpendicular to it).

So there is a stress which relaxes as the polymer wriggles out of its old tube into a new equilibrium. The stress is related to the amount of polymer remaining in the original tube. This simple picture when quantified gives good agreement with experiment.

4 CONCLUSION

We have noted that food materials exhibit a wide range of molecular complexity and cross-link type. Theories exist which allow us to begin to resolve mechanical effects, gel types, and junction (cross-link) behaviour. In the following chapter, theories are applied to simple and composite gelling systems.

2 The Application of Network Theory to Food Systems

A. H. CLARK

Unilever Research, Bedford

1 INTRODUCTION

Aqueous solutions of biopolymers can often be converted to gels by thermal or chemical means. This process of network formation is of value in food applications,[14] as it involves a drastic change in rheological behaviour and immobilization of a considerable volume of water. Of the two gelling procedures referred to, the thermal mechanism is usually the more convenient, particularly in food processing, and the present paper considers thermally-induced gelation only. Some of the conclusions reached, however, and the experimental approaches described should be relevant to the case of chemically-induced cross-linking as well.

As a phenomenon, the thermal setting of biopolymer solutions to form gels can be separated into two distinct classes of process, i.e. heat-setting and cold-setting. The aggregation of thermally unfolded globular proteins[1,3,4,27] including proteins from muscle[24] is a good example of heat-set gelation, whilst the cross-linking of gelatin, a fibrous protein[8,20] and various types of polysaccharide,[17,18,21] via disorder-to-order transitions induced by cooling, is typical of the cold-setting process. Not only are these two classes of

FOOD STRUCTURE AND BEHAVIOUR
ISBN 0-12-104230-8

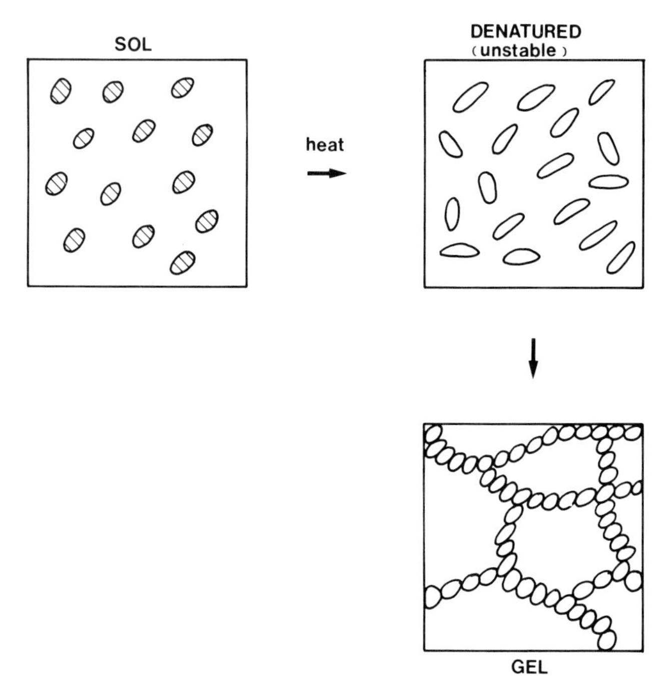

Fig. 1 Schematic representation of the formation of a corpuscular network from thermally-unfolded globular protein molecules. In this model the polymers enter the network in a compact form as the unfolding process is rarely drastic enough to generate anything approaching a random coil state

biopolymer gel different in their origins, but they are also fundamentally different at a molecular level, for as is indicated schematically in Figs 1 and 2, the globular protein case involves aggregation of polymers still in corpuscular form whilst the cross-linking of polysaccharides and gelatin involves extended statistical coils associated along their lengths via formation and interaction of ordered chain conformations such as single and multiple-stranded helices.

Whichever process is involved, however, heat-setting or cold-setting, and whichever type of network is formed, cross-linking of macromolecules is only part of the requirement for stable gel formation, as forces are also necessary which oppose attraction and tend to keep polymers in solution. Whilst solvation may partly contribute to this process, the most important factor appears to be electrostatic repulsion, for as is evident in the globular protein example, homogeneous gels form only at pHs where the protein is substantially ionised. Adjusting pH towards the isoelectric value, or raising ionic strength, causes protein gels to become progressively more and more

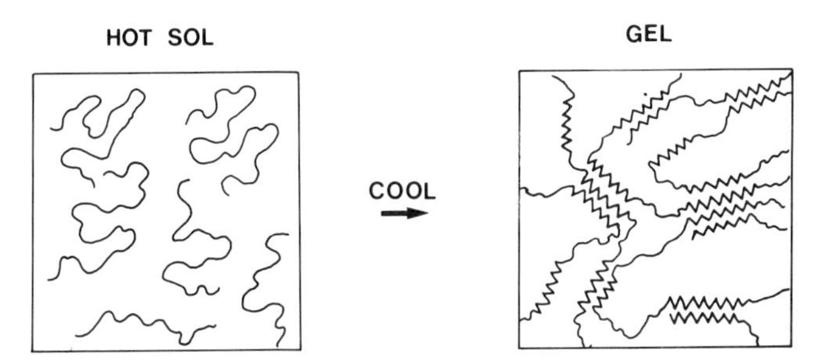

Fig. 2 Schematic representation of the formation of an association network as a hot solution of a random coil polymer is cooled. This behaviour is typical of gelatin and certain polysaccharides, and cross-linking involves formation and aggregation of single or multi-stranded ordered chain conformations

turbid and water-releasing, and where turbidity and syneresis also occur in polysaccharide gels they appear to have similar origins. Overall, it is clear that the phase behaviour of aqueous gels, and its relationship to gelling conditions, are important aspects of biopolymer gelation, as they have a profound influence on macroscopic attributes such as gel texture, appearance, and water-binding capacity.[13]

Whilst substantial phase separation is only sometimes a property of single-component gels, it is almost always found in the case of composite gels based on two unlike biopolymer components such as a globular protein and a polysaccharide. Mixed aqueous gels based on these polymers can readily be made[5,6] by processes involving heating and cooling, but because of the fundamental incompatibility of these components, microscopically phase-separated materials result which scatter light strongly. Both light and electron microscopy have been used in the study of such systems,[5,6] and reveal complex microstructures involving dispersion of one polymer gel phase in the other (see, for example, Fig. 3), and phase inversion at particular mixture compositions. A parallel is found with the microstructures of composites based on mixtures of synthetic polymers,[16] and this has prompted theoretical treatments of mixed gel rheology[6] based on the gelling properties of individual components and simple additivity rules. In the present paper the characterisation of individual biopolymer gelling systems necessary for this modelling is described, and the way such data can be used to understand composite rigidity is explained. In the end it is contended that the rheological properties of mixed biopolymer gels may more often depend on component incompatibility and phase separation than on the converse situation of specific "molecular synergism", but it is recognized that the systems described are comparatively simple and serve only as models for the more complex types of mixed gels encountered in real food applications.

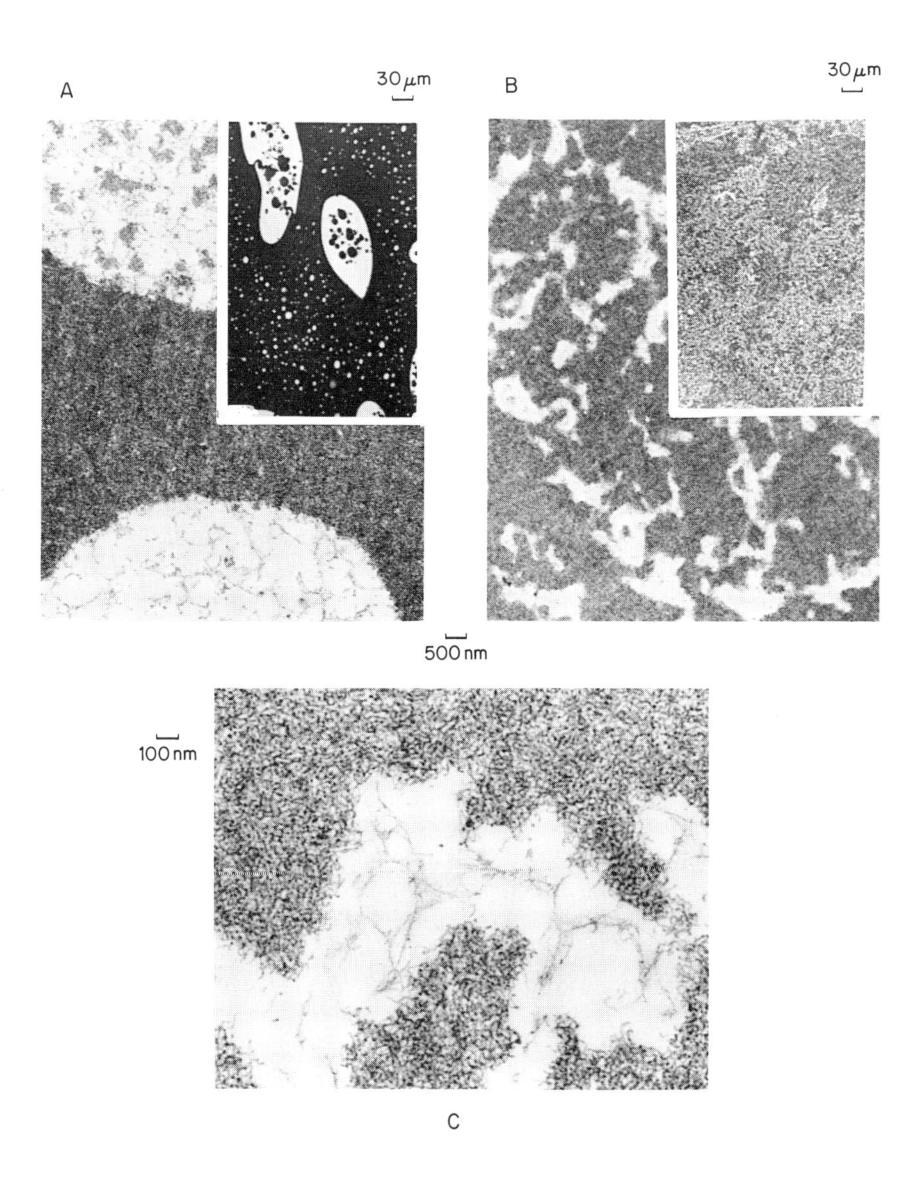

Fig. 3 Transmission electron and light micrographs (inserts) indicating the complex microstructures of mixed biopolymer gels. In this example the biopolymer components are agar (a polysaccharide) and bovine serum albumin (BSA) a globular protein. The darkly-staining material is the protein (present at 10% w/w concentration) and the lighter areas contain agar (1% w/w). Figure 3*A* shows a coarse phase-separated structure produced by gelling the protein first; 3*B*, a more homogeneous structure produced by gelling the agar first; and 3*C*, a contrasting view of the BSA and agar networks at higher electron microscope magnification. See [5].

2 THEORETICAL ASPECTS OF THE GELATION PROCESS

Detailed knowledge of the molecular structure of a gelling biopolymer system is likely to be of little practical use unless it can be linked to some macroscopic aspect of the gelation process, such as the mechanical behaviour of the gel, its water-holding capacity or its ability to scatter light. Of these properties, the mechanical behaviour is probably the most important, as the most prominent feature of the gelation phenomenon is the transformation from liquid to solid as molecules are cross-linked to form a three-dimensional network. Gelation is a consequence of branched molecular polymerization and indeed statistical theories[9,26] of molecular polymerization have long been known which include gel formation as a special case. Although these theories were developed for condensation of simple polyfunctional organic molecules, and cross-linking of preformed chains, in situations where the monomeric species often constituted the solvent, such early statistical models and a more recent elaboration involving cascade theory[10,11] are still relevant to the case of biopolymers cross-linking in aqueous media, and provide an interesting insight into the network-building process. For example, if cross-linking is random, and each biopolymer molecule has f equivalent sites available for bond formation, classical gel theory[9] predicts that a critical fraction of these must react before the first vestiges of a three-dimensional network can appear. This fraction α_c is related to f by the equation,

$$\alpha_c = 1/(f-1) \tag{1}$$

and at degrees of cross-linking α less than this, only aggregates of limited size can exist. When, however, after a certain amount of reaction time, usually referred to as the gel time, the critical value α_c is approached, one molecule outstrips all the others in size (by many orders of magnitude) and becomes the "gel fraction". At this stage the solution viscosity becomes extremely large, and as M_w, the weight-average molecular weight, diverges to infinity the system is converted to a soft viscoelastic solid. As cross-linking proceeds further (i.e. as α exceeds α_c) the shear modulus of the gel increases, and at the same time the "sol fraction" or collection of finite molecular aggregates gets smaller. Eventually, this fraction disappears altogether, and further increase in gel strength requires intramolecular cross-linking, i.e. cross-linking within the gel molecule itself. Finally, at long times the modulus ceases to increase significantly as further cross-linking becomes very slow, or a position of equilibrium between bond-making and breaking (if reversibility is possible) is achieved.

This last situation of a limiting modulus at long times is particularly interesting and important in the biopolymer case, as for many biopolymer systems certain concentrations are found below which gel formation (as indicated by a detectable shear modulus) is impossible in a reasonable amount of time. This effect, which is best demonstrated by plotting gel time on a logarithmic scale against biopolymer concentration, and showing a divergence to infinity at some critical concentration value C_0, means that below this concentration α_c, the critical degree of cross-linking, cannot be realised in a finite time by the kinetic processes operative. In situations where physical attractions are involved, reversibility of the cross-linking process may provide a simple explanation for this phenomenon, but whatever its physical origins, critical concentration behaviour is an experimental feature of biopolymer network formation in solutions and one which is not considered in classical statistical descriptions of the gelation process.

The classical description of gelation is limited in other ways too particularly in relation to the topic of phase separation in gels, for, in the theoretical discussion so far, it has been tacitly assumed that gelling systems are physically homogeneous at all polymer concentrations and degrees of cross-linking. As in the case of critical concentration, however, complex phase behaviour is a common aspect of biopolymer network formation,[3] and one which is ignored in classical statistical theories. Where gelling systems become substantially inhomogeneous, they also become statistically inhomogeneous, and it seems that α must be thought of as dividing into two components, one referring to denser parts of the network, and the other to less dense regions. It must be added, however, that an alternative approach to biopolymer gelation based on statistical mechanics has appeared recently[7] which attempts to treat gelling in solution, and which predicts critical concentration behaviour, and physical inhomogeneity. As yet the practical utility of this theory in the biopolymer gel case has not been fully established, and in what follows the classical theory will still be pursued, high gel homogeneity being assumed for situations where single polymers are considered, and reversibility of cross-linking being invoked to explain the critical concentration aspects of the gelling process.

3 KINETIC ASPECTS OF BIOPOLYMER GELATION BOTH BEFORE AND AFTER THE GEL POINT

The theoretical description of gelation just outlined allows the process to be divided into two parts, i.e. aggregation prior to the gel point ($\alpha < \alpha_c$), and network formation afterwards ($\alpha \geqslant \alpha_c$). Experiments designed to follow gelation kinetics can be categorized in the same way, and in this section some results for biopolymers are described.

Aggregation in solution prior to the gel point can be followed using techniques such as viscometry, osmometry, and light scattering, while spectroscopic techniques such as optical rotation can also be of value. Tombs,[27] for example, has reported studies of the heat-induced aggregation of the globular protein bovine serum albumin (BSA) in which the number average molecular weight M_n was followed by osmometry, and Kratochvíl and coworkers[15] have used light scattering to study a similar process for human serum albumin (HSA). In this latter investigation the weight average molecular weight M_w was followed as a function of time, and a theoretical analysis of the data was made using a kinetic model based on the statistical theory of gelation, and a second order rate process describing coupling of cross-linking sites. This equation (which does not allow for reversibility, and can only be used to describe the early stages of association) is identical to one subsequently used by Parker and Dalgleish[19] to describe casein aggregation, and can be written in the form,

$$M_w = \frac{M(1 + 2fv_0kt)}{(1 - f(f-2)v_0kt)} \tag{2}$$

where M is the monomer molecular weight, f (>2) is the number of equivalent binding sites per monomer, v_0 is the concentration of monomer initially present, k is the rate constant, and t the time. Unlike the better known Smoluchowski equation[25] developed much earlier to describe particle aggregation, the Kratochvíl equation predicts gel formation at a finite time, for M_w diverges to infinity at $t = 1/f(f-2)v_0k$. Using weight-average molecular weight versus time data, and a curve-fitting procedure based on equation 2, Kratochvíl *et al.* determined a best value for f of 2.05 for HSA aggregation, the fractional functionality being interpreted as indicating a polydispersity in the number of binding sites available per unfolded protein monomer. Though the Kratochvíl approach is a general one, apart from the work of Parker and Dalgleish,[19] it does not appear to have been applied very often, and indeed it appears that studies of the incipient gelling behaviour of biopolymers are not that common in the literature.

Studies of aggregation through the gel point are perhaps even fewer in number, however, as it is no longer possible to apply light scattering methods rigorously, viscosity ceases to be a measureable quantity, and in some cases even spectroscopic methods fail on account of turbidity. In fact, when a gel is in process of forming the most useful quantity to measure is the viscoelastic shear modulus $G^*(\omega)$ in the form of its storage and loss components, $G'(\omega)$ and $G''(\omega)$. These can be measured at a particular frequency ω, using a torsion pendulum, or over a range of frequencies using a mechanical spectrometer, and using either technique the development of the modulus can be monitored as a function of time (see, for example, Fig. 4). Experiments of this kind have been successfully performed by Richardson and Ross-Murphy[22] for BSA aggregation at various protein concentrations, and at

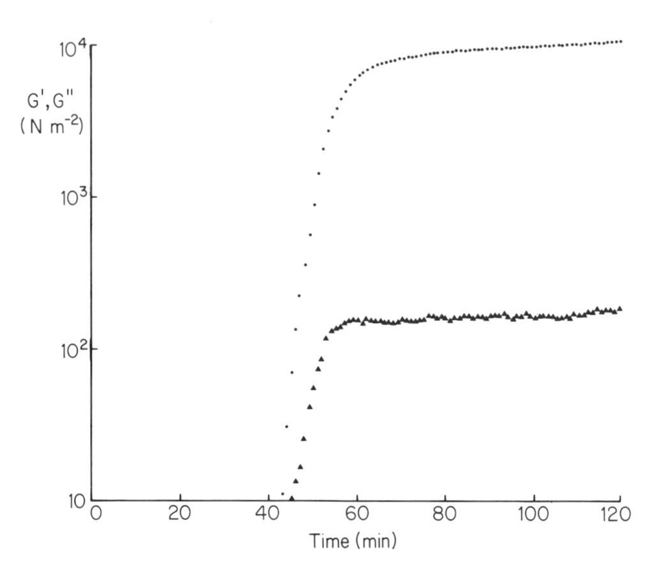

Fig. 4 Typical time dependence of the storage and loss components (G', G'') of the viscoelastic shear modulus of a gelling biopolymer solution. The lag period, or gel time, required for development of a measureable modulus is clearly indicated, as is the rapid increase in modulus soon after the gel point, and its ultimate plateauing off at long times

constant heating temperatures, and the effect of concentration on gel time (i.e. the time lag before G' becomes measurable) and the form of the G' versus time relationship have been discussed. Detailed interpretation of such data is difficult, however, without an independent measure of $\alpha(t)$ the degree of cross-linking, and whilst in the synthetic polymer case this may be possible experimentally by monitoring unreacted functional groups, it is much less easy to establish for biopolymers interacting through physical forces. In their solution to this problem Richardson and Ross-Murphy used optical rotation measurements both before and after the gel point to estimate $\alpha(t)$, though in doing so they had to assume that cross-linking of BSA molecules and optical rotation change were in direct proportion. Using $\alpha(t)$ determined in this way, G' became known as a function of α, and a functionality f was determined which was very similar to the Kratochvíl result[15] for HSA. In addition, assuming this G' versus α relationship, it was possible to employ the methods of cascade theory[10, 11] to comment on the mechanical characteristics of the BSA networks formed, for in the language of the cascade formalism, G' should relate to α via the equation,

$$G' = aRT\{Nf\alpha(1-v)^2(1-\beta)/2000\} \quad (3)$$

in which,

$$\beta = (f - 1)\alpha v/(1 - \alpha + \alpha v) \tag{4}$$

and v is the so-called "extinction probability" defined by the relationship,

$$v = (1 - \alpha + \alpha v)^{f-1} \tag{5}$$

In equation 3, N is the concentration of monomer species in moles per dm^3, R is the gas constant, T the absolute temperature of the gel, f the number of equivalent cross-linking sites per particle, and "a" is a measure of the number of RT units contributed to the modulus by each mole of load-bearing chains in the network. Using their estimate of f, Richardson and Ross-Murphy[22] fitted equation 3 to the modulus versus α data, and showed that a good fit required "a" to be much greater than the value of unity anticipated for an ideal rubber. This, and a tendency for G' to fall with increasing temperature, were considered as evidence for non-rubberlike behaviour (i.e. an enthalpically rather than entropically determined modulus) and this conclusion was considered reasonable in view of the likely complexity of elastically-active network chains based on partially-unfolded[4] BSA molecules.

Though an approach similar to the one described has also been applied in the case of gelatin gelation,[20] it seems that the relationship between modulus and time has been little investigated for biopolymer gels, probably because of the uncertainties and difficulties inherent in methods available for experimental determination of $\alpha(t)$. The kinetic approach to biopolymer gelation should be pursued, however, for one of the most intriguing aspects of this phenomenon is whether it is essentially an irreversible process, or whether, where physical forces are concerned, cross-linking finds its way to an equilibrium (not necessarily the thermodynamically most stable one) where cross-link formation is opposed by bond breaking. So far, for most systems, this question has not been satisfactorily answered, and as it has a considerable bearing on the subject to be addressed next, i.e. the relationship between limiting gel rigidity and biopolymer concentration, this is unfortunate.

4 THE CONCENTRATION DEPENDENCE OF THE GEL MODULUS

As was stated in the Introduction, characterization of the modulus versus concentration behaviour of a gelling biopolymer system is required if its contribution to composite gel rigidity is to be rationalised. In the present section, therefore, the relationship between G' and concentration for a single biopolymer species is the issue in question. Before considering theoretical aspects of this relationship, however, it is necessary to define the modulus to be measured, for as has been said G' changes with time as cross-linking proceeds, and depends on the measurement frequency. In fact where experimental data are discussed in this section it is to be assumed that frequency has been fixed at a value dictated by the measurement apparatus, or by the timescale of interest to the experimentalist, and the modulus referred to will be a limiting value measured only after curing has gone on long enough to ensure very little further increase of G' with time. The measurement itself will have been made for the gel after its temperature has been adjusted to a suitable value, though there is no reason in principle why the modulus should not be quoted at any temperature including that at which cross-linking originally took place. In the present case, G' measurements will be quoted at 25 °C and at a frequency of ~ 0.2 Hz, and will be presented for three types of biopolymer gel. In each case a series of biopolymer concentrations will be considered, and a critical concentration will be demonstrated for each system below which G' was negligibly small even after prolonged heating (or cooling) of solutions.

From the nature of the experimental procedure just described it is clear that any theoretical interpretation of its consequences will require a model for the kinetics operative at long times, and this might involve assumption of a dynamic equilibrium, or of an irreversible cross-linking process progressing very slowly. From what was said earlier about kinetics and gelation, it is likely to be difficult to establish such a model by kinetic experiments alone, and in the end the assumptions made may have to be judged in relation to the success they achieve in producing a satisfactory fit to modulus versus concentration data. In this section, the equilibrium hypothesis is explored, and is found to work well in practice, even though initially its foundations as a description of biopolymer gel cross-linking might seem somewhat shaky.

Experiments of the type outlined above, and interpretations based on the equilibrium hypothesis have already appeared in the literature, and have achieved some success. Bikbov *et al.*[2] for example, have published data for soy protein gels, and have shown how a theoretical equation derived by Hermans[12] could be fitted to over thirty data points, and could predict critical concentration behaviour quite naturally. In addition, Richardson and Ross-Murphy[23] have fitted similar concentration data for BSA gels and have concluded that the Hermans' theory has much to offer in the strong gel case, despite its origins as a theory derived for weak polymer-polymer interactions. The only difficulty Richardson and Ross-Murphy experienced was that whilst Hermans' formula was derived for molecules having large f only, according to their previous work on the incipient gelling behaviour of BSA solutions,[22] the BSA functionality was likely to be small, and in principle, Hermans' function ought not to have been so readily applicable to its case.

In an attempt to clarify this point by establishing the sensitivity of the concentration dependence of the gel modulus to the functionality of the polymerizing species, the present author has reformulated the original Hermans' approach using the more convenient equations of cascade theory,[10,11] and has derived a corresponding expression for the G' versus concentration relationship valid for any f greater than two. The argument employed was the following one. Let there be N moles of biopolymer per dm^3 of solution, or alternatively, let the concentration be expressed as C weight percent. If M is the molecular weight,

$$N = 10\, C/M \tag{6}$$

If there are f equivalent sites per molecule potentially available for bonding to other molecules, then Nf is the concentration of such sites originally present in solution. If, after a certain period of cross-linking, it is assumed that equilibrium has become established between bond-making and breaking (Hermans' assumption), then α, the fraction of sites which have reacted, can be written in terms of an equilibrium constant K as follows,

$$\alpha/Nf(1-\alpha)^2 = K \tag{7}$$

Here K is in units of dm^3 per mole and equation 7 is derived by assuming a second-order forward reaction and first-order back reaction for the association of reactive sites. It should be noted that concentrations of aggregated species such as dimers, trimers etc. are not discussed explicitly in this treatment, the kinetic argument being couched in terms of the concentration of reactive sites rather than particles.

Equation 7 can be solved for α, the result being,

$$\alpha = 1 + \frac{1}{2P}\left(1 - \sqrt{4P+1}\right) \tag{8}$$

where

$$P = NfK = 10fKC/M \tag{9}$$

When a gel is just able to form, C has the critical value C_0, and α is given by equation 1, i.e. $\alpha = 1/(f-1)$. If this result is substituted in equation 8, C_0 may be written in terms of f, K and M as,

$$C_0 = \frac{M(f-1)}{10Kf(f-2)^2} \tag{10}$$

Here C_0 is the critical concentration in weight percent and is proportional to the molecular weight and inversely related to the affinity for cross-linking as measured by K, and the number of binding sites f. From the result for C_0

given in equation 10 it is clear that P can be written in the form,

$$P = \frac{(f-1)C}{(f-2)^2 C_0} \tag{11}$$

and this means that α is always a function of f and C/C_0 only, and does not depend explicitly on M and K.

With α established, it is now possible to use the methods of cascade theory to calculate the number of elastically active network chains per mole of biopolymer present. The result is,

$$n = f\alpha(1-v)^2(1-\beta)/2 \tag{12}$$

where v and β are defined by equations 4 and 5 given previously. It follows that the number of moles of such chains per cm^3 of gel is given by equation 12 multiplied by $N/1000$, and hence the equilibrium modulus G_e is obtained by multiplying this result by aRT as discussed earlier (equation 3). The equilibrium modulus G_e is the hypothetical modulus value which would be measured at extremely low frequencies, but in the present work it will be assumed that G' measured at 0.2 Hz is a close approximation to it, i.e. it is assumed that all physical cross-links of importance are stable over a timescale much greater than one second. By combining equation 3 with equations 6 and 10, an expression for the shear modulus G (equivalent to G_e or G' at 0.2 Hz) can be written as,

$$\frac{1000KG}{aRT} = \frac{C}{C_0}\left\{\frac{(f-1)\alpha(1-v)^2(1-\beta)}{2(f-2)^2}\right\} \tag{13}$$

and since α (and hence v and β) depend on f and C/C_0 only (see equation 11) then equation 13 takes the form of a universal function in which G, divided by a scaling constant, is expressed as a function of C, also divided by the constant C_0. Features peculiar to any individual polymer system such as M and K are contained solely in the scaling constants, but it is important to note that an explicit dependence on f remains, and cannot be removed completely. Dependence of the master curve on f can be reduced, however, by multiplying both sides of equation 13 by $(f-1)(f-2)$ to give the convenient expression,

$$\frac{1000KG(f-1)(f-2)}{aRT} = \frac{C}{C_0}\left\{\frac{(f-1)^2\alpha(1-v)^2(1-\beta)}{2(f-2)}\right\} \tag{14}$$

and indeed, the factor of 1000 appearing on the left-hand side of this equation can be omitted if it is understood that the units of G, K and R are to be chosen such that KG/RT is dimensionless.

Master curves calculated for various values of f between 3 and 1000 are shown in Fig. 5, the $f = 1000$ model being similar (but not identical) to that

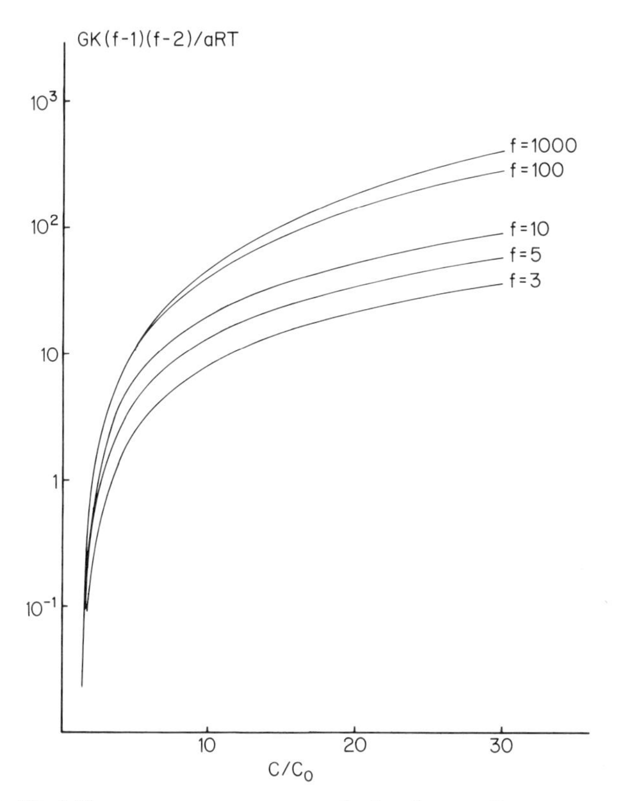

Fig. 5 Modified Hermans master curves calculated according to equation 14 for various choices of functionality *f*. Units for *G*, *K* and *R* have been chosen compatibly so that the ordinate axis is dimensionless, as is the abscissa

proposed by Hermans[12] using Flory's gel theory.[9] Use of equation 14 to fit real data involves choosing a value for f and determining best values for the scale factors C_0 and $aRT/1000K(f-1)(f-2)$ by least-squares fitting. This in turn amounts to determining best values for K and "a", for a measured temperature T and known molecular weight of the associating species, M. Fits for different choices of f are compared, and often it is possible to select a value for f which gives a best overall fit. Knowing the best-fit scale factors it is then possible to use equation 14 in reverse to reduce experimental data to master curve form, a procedure which is usually informative as it can be used to compare data sets for different polymer systems, or for a given polymer gelled under different experimental conditions.

The applicability of equation 14 to experimental data can readily be demonstrated, and some examples of its use in the biopolymer field are shown

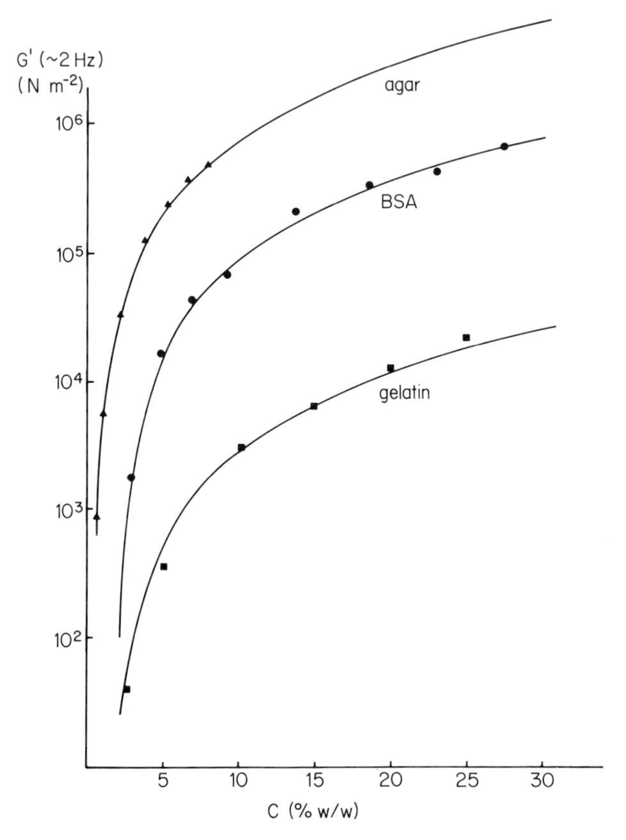

Fig. 6 Fits of the $f = 1000$ master curve to modulus (G') versus concentration data for agar (▲), BSA (●) and gelatin (■) gels. The heat-set BSA gels were made by prolonged heating at 95 °C followed by cooling at 25 °C, and the cold-set agar and gelatin gels by prolonged cooling at 25 °C

in Figs 6 and 7. These show experimental G' versus concentration data for heat-set globular protein gels formed from BSA, and cold-set gels from gelatin[6] and the polysaccharide agar.[6] Figure 6 shows fits using a high-functionality approach ($f = 1000$) virtually equivalent to the original Hermans' treatment, and Fig. 7 shows what happens when f is reduced to three, a value apparently more realistic in the BSA situation.[22] In general, the high functionality model fits a little better to all sets of data, but differences are not enormous, and are probably within the limits of uncertainty imposed by the assumptions of the cascade theory, and the equilibrium approach. It appears that f cannot be estimated with any certainty from such fitting procedures alone, but it emerges very clearly from details of the fits carried

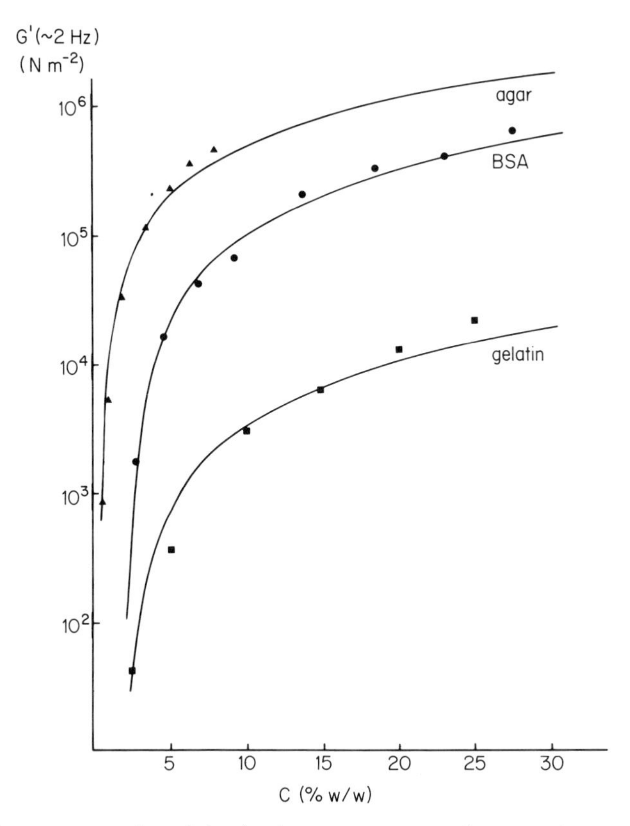

Fig. 7 Least-squares fits of the $f = 3$ master curve to the same data as in Fig. 6

out, that when f is small, "a" is much larger than the rubber theory estimate of one, and this is consistent with Richardson and Ross-Murphy's conclusions described earlier, and with the structured and complex nature of the elastically-active strands expected in biopolymer networks. The parameters f and "a" are highly correlated it seems, so that to determine f, "a" must be known from some other source, and vice versa.

Whatever may be argued about the theoretical basis of the equilibrium model, and about the physical significance of f, "a" and K, it is certain that equation 14 and the original Hermans' function, give good fits to experimental data, and since in the treatment of composite systems to be described in the next section, this property is required[6] for purposes of interpolation and extrapolation, the equilibrium model may be considered of value. It is to be hoped, however, that in the long run it will amount to more than just a function for fitting data, and that its use will allow the viscoelastic properties of biopolymer gels (in the linear region, that is) to be more clearly understood as data become available for more and more systems and sets of gelling conditions.

5 A THEORETICAL MODEL FOR THE LINEAR VISCOELASTIC BEHAVIOUR OF COMPOSITE BIOPOLYMER GELS

In the study of binary composite systems based on synthetic polymers the most elementary approach to modulus prediction is through simple additivity formulae[16] such as,

$$G_C = \phi_X G_X + \phi_Y G_Y \tag{15}$$

and,

$$1/G_C = \phi_X/G_X + \phi_Y/G_Y \tag{16}$$

where G_C is the shear modulus of the composite, and G_X and G_Y are the moduli of its two components considered as individual systems present at phase volume fractions ϕ_X and ϕ_Y. The derivation of equations 15 and 16 is in fact quite simple and is carried out by assuming two extreme cases for the distribution of stress and strain within the composite. Thus, equation 15 refers to an upper limit for the composite modulus corresponding to a situation where the strain field is uniform throughout the material and the stress is discontinuous (isostrain model), and equation 16 describes a lower bound appropriate to a situation of uniform stress and discontinuous strain (isostress model). Figure 8 shows thesc bounds for a material made from a comparatively weak component X and a stronger component Y. As the phase volume fraction of Y increases, a phase inversion is expected with Y replacing X as the supporting phase, and when this happens it is anticipated that the modulus of the system (broken line) will shift from lower to upper bound behaviour. In practice, in the field of synthetic polymer composites, the exact behaviour of the modulus as ϕ_Y varies is usually more complicated, and more elaborate phenomenological models based on combinations of equations 15 and 16 have been proposed.[16] The simple upper and lower bound limits remain useful, however, as they define an area of modulus versus composition space in which experimental results should lie if the rigidity of the material is determined by simple phase separation.

When attention is shifted from the synthetic polymer case (without a solvent) to mixed aqueous biopolymer gels, the same approach can be adopted, but the extra solvent component (water) complicates the issue, for when phase separation occurs during gelation the water present becomes

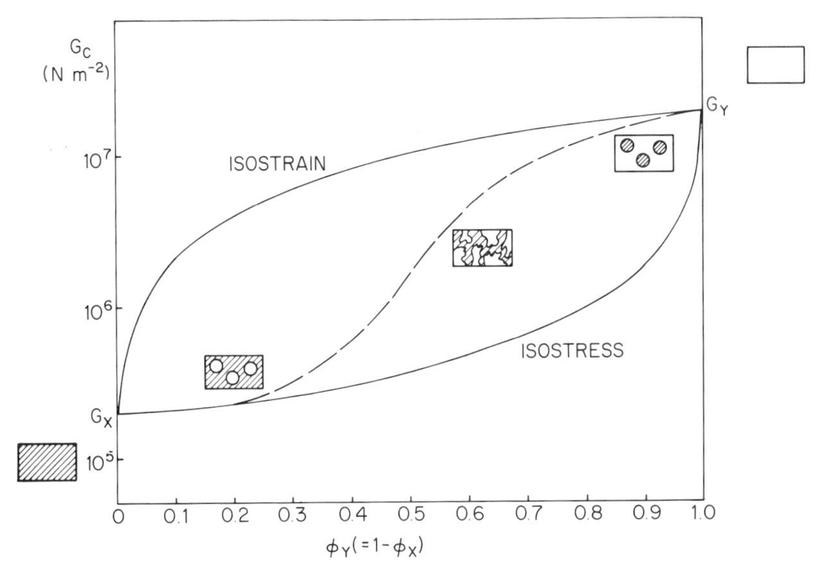

Fig. 8 Isostrain and isostress bounds (equations 15 and 16) for a hypothetical composite system containing polymers X and Y and no solvent. Y is considerably more rigid than X, and phase inversion occurs at some intermediate stage as Y is added to X. The broken line indicates a possible variation of the actual composite modulus with composition, i.e. a shift from lower to upper bound behaviour as the more rigid polymer becomes the supporting phase

partitioned between components in an unknown way. It is for this reason that it is necessary to study modulus versus concentration relationships for the pure components, for in the aqueous composite case, G_X and G_Y are not fixed quantities, but are functions of the local concentrations of X and Y, present in any particular situation, and produced by the phase separation process.

In a recent publication[6] on the mechanical properties of mixed aqueous gels based on the polysaccharide, agar and the protein gelatin, Clark *et al.* have described a way in which the simple additivity model (equations 15 and 16) can be adapted to cope with the three-component case. The argument was as follows. If x and y are weight percentages of X and Y in solution, and $w = 100 - x - y$ is the mass of associated water, then the nominal concentrations of X and Y are just x and y. When phase separation occurs, however, αw of the water becomes associated with X, and $(1 - \alpha)w$ with Y, where α is, as yet, an unknown fraction. In this situation the local concentrations of X and Y in the two phases, and those effective in generating gel strength become,

$$C_X^{\text{eff}} = 100x/(x + \alpha w) \tag{17}$$

and

$$C_Y^{\text{eff}} = 100y/(y + (1 - \alpha)w) \quad (18)$$

If the densities of the two phases are now assumed equal (since they are generally dilute aqueous gels), then the volume fractions of regions X and Y are given by the equations,

$$\phi_X = (x + \alpha w)/100 \quad (19)$$

and

$$\phi_Y = (y + (1 - \alpha)w)/100 \quad (20)$$

In equations 17 to 20, α appears as a variable fraction (not to be confused with α the fraction of reacted cross-linking sites discussed earlier) which depends both on the composition as defined by x and y, and on the intrinsic relative affinity of the two polymers for water. An attempt may be made to separate these factors by introducing a parameter p which is assumed to be independent of x and y, and which is introduced to describe the intrinsic relative powers of attraction of X and Y for water. In terms of p, α can be written as,

$$\alpha = px/(px + y) \quad (21)$$

and clearly $p < 1.0$ implies that Y is more hydrophilic than X. It is likely that p can be measured for a system by studies of the osmotic pressure properties of the individual components, but for the moment, if equation 21 is assumed, equations 17 to 20 can be rewritten as,

$$C_X^{\text{eff}} = 100(px + y)/(100p + (1 - p)y) \quad (22)$$

$$C_Y^{\text{eff}} = 100(px + y)/(100 - (1 - p)x) \quad (23)$$

$$\phi_X = x(100p + (1 - p)y)/100(px + y) \quad (24)$$

$$\phi_Y = y(100 - (1 - p)x)/100(px + y) \quad (25)$$

It follows that, for all x and y, provided that p is known or can be guessed for the polymer pair and solvent in question, bounds for G_C can be calculated from equations 15 and 16, and the phase volume and concentration information in equations 22 to 25. The concentrations C_X^{eff} and C_Y^{eff} determine the values G_X and G_Y to be used in equations 15 and 16, as these moduli can be estimated by interpolating experimental G (in most cases G' is what is measured) versus concentration data in the manner discussed in the last section. Using this procedure, upper and lower bounds can be calculated for any set of composition variables x and y, and for any choice of p.

As an example of this procedure, Fig. 9 shows three sets of bounds

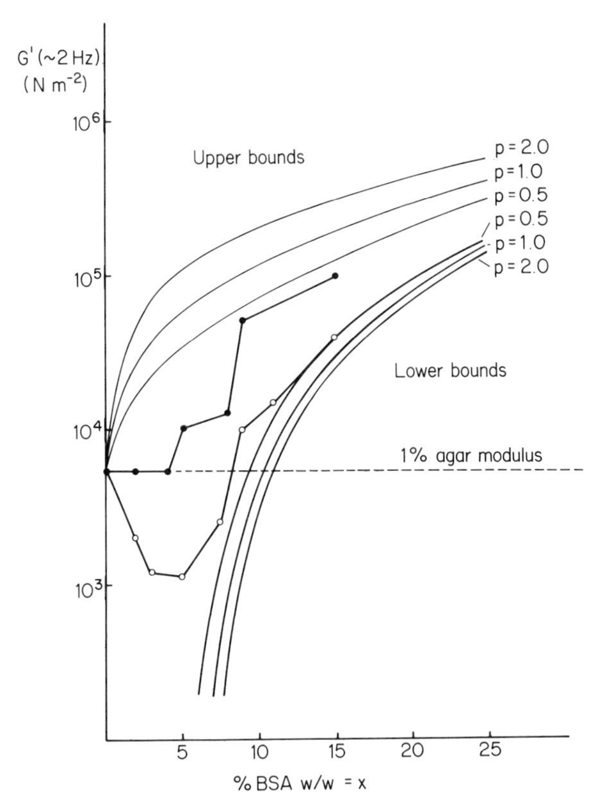

Fig. 9 Experimental modulus versus composition data for agar/BSA composite gels plotted in relation to theoretical upper and lower bounds calculated as described in the text for $p = 0.5$, 1.0 and 2.0. The nominal agar concentration was fixed at 1% w/w, and the nominal BSA concentration (x) allowed to vary. In one gel series the protein was gelled first (●) in the other the agar (○). See [5].

calculated for a co-gelling system based on agar and the globular protein BSA. The agar concentration y has been fixed at 1% w/w, and the BSA concentration allowed to vary in the range 0 to 25% w/w. Three values of p have been assumed, namely 0.5, 1.0 and 2.0. Because of the more hydrophilic character of the agar component, the 0.5 value might seem the most realistic, but in fact, at present, no precise estimate for p is available from experiment.

In Fig. 9, experimental data points[5] are also shown for two series of agar-BSA gels made under different thermal conditions (see also Fig. 3). As may be seen, all observations lie within the calculated bounds, one series of points following the lower bound behaviour quite closely at higher concentrations, and the other doing the reverse. From the point of view of specifying bounds the model appears to work well, and the dependence of the results on choice of p does not appear to be crucial.

Questions about the detailed dependence of G on x are of course more difficult to answer, though some considerable discussion of this matter has been given in the work on agar-gelatin systems.[6] Ideally, it would be expected that agar-protein composites would follow upper bound behaviour prior to phase inversion (when the agar gel is the supporting phase) and lower bound behaviour afterwards when the weaker protein gel takes over. Though exact inversion points are not known for the agar-BSA systems in question, they are believed to occur close to $x = 5\%$ w/w. For one gel series, the lower bound behaviour certainly occurs after this point, but this is not so for the other, and in neither case is upper bound behaviour observed at small x. As has been argued in the agar-gelatin case[6] a number of factors influence the exact behaviour of the modulus in these systems and these include the order in which the components gel, the positions of phase inversion points, and the exact morphologies of the phase-separated structures produced. Whilst some success in explaining these matters has been achieved for agar-gelatin, the behaviour of the agar-BSA data in Fig. 9 has not been fully explained. It can be concluded, nonetheless, that in broad terms, the behaviour of this and other systems based on unlike polymers can be rationalised satisfactorily in terms of the polymer incompatibility concept and the phase separation property which is its natural consequence.

6 CONCLUSIONS

The subject of biopolymer network formation has been reviewed, and physical aspects of the process considered. A model has been developed to describe shear modulus versus concentration data based on the assumption of a cross-linking equilibrium at long times. Though the validity of this model remains to be fully established, it is clearly of practical value in the study of mixed gel rheology, for it provides a means of interpolating and extrapolating experimental data. Such procedures allow upper and lower bounds to be produced for the shear modulus of composite gel systems, and though the exact modulus versus composition behaviours of these gels are as yet obscure, the method should enable effects due to specific molecular interactions to be identified and separated from properties which result simply from physical phase separation. It must be admitted, however, that in terms of the applicability of this approach to real food systems, tests have so far been confined to model mixed gels only, and its value in allowing the properties of genuine food materials to be understood and predicted remains to be demonstrated. The degree of success which has been achieved for systems such as the agar-gelatin composites allows guarded optimism, however, and suggests that the method described will at least provide an interesting avenue for future research to those interested in the quantitative behaviour of food substances containing gelling biopolymer components.

Acknowledgements

In conclusion, the author would like to thank his colleagues Dr S. B. Ross-Murphy and Mr R. K. Richardson for many useful discussions of polymer gelation and for provision of experimental facilities and data. He also thanks Mr G. Robinson and Mr J. M. Stubbs for microscope images of the mixed gel systems, and Mrs L. A. Linger for technical assistance in the preparation of Figures.

References

1. Barbu, E. and Joly, M. (1953). *Faraday Discuss. Chem. Soc.* **13**, 77–93.
2. Bikbov, T. M., Grinberg, V. Ya., Antonov, Yu. A., Tolstoguzov, V. B. and Schmandke, H. (1979). *Polymer Bull.* **1**, 865–869.
3. Clark, A. H., Judge, F. J., Richards, J. B., Stubbs, J. M. and Suggett, A. (1981). *Int. J. Peptide Protein Res.* **17**, 380–392.
4. Clark, A. H., Saunderson, D. H. P. and Suggett, A. (1981). *Int. J. Peptide Protein Res.* **17**, 353–364.
5. Clark, A. H., Richardson, R. K., Robinson, G., Ross-Murphy, S. B. and Weaver, A. C. (1982). *Prog. Fd Nutr. Sci.* **6**, 149–160.
6. Clark, A. H., Richardson, R. K., Ross-Murphy, S. B. and Stubbs, J. M. (1983). *Macromolecules* **16**, 1367–1374.
7. Coniglio, A., Stanley, H. E. and Klein, W. (1979). *Phys. Rev. Lett.* **42**, 518–522.
8. Eagland, D., Pilling, G. and Wheeler, R. G. (1974). *Faraday Discuss. Chem. Soc.* **57**, 181–200.
9. Flory, P. J. (1941). *J. Am. Chem. Soc.* **63**, 3083–3100.
10. Gordon, M. (1962). *Proc. Roy. Soc. (London) Ser. A* **268**, 240–259.
11. Gordon, M. and Ross-Murphy, S. B. (1975). *Pure Appl. Chem.* **43**, 1–26.
12. Hermans, J. (1965). *J. Polymer Sci., Part A* **3**, 1859–1868.
13. Hermansson, A.-M. (1982). *J. Fd Sci.* **47**, 1965–1972.
14. Kinsella, J. E. (1976). *CRC Crit. Rev. Food Sci. Nutr.* 219–280.
15. Kratochvíl, P., Munk, P. and Sedlacek, B. (1961). *Coll. Czech. Chem. Commun.* **26**, 2806–2811.
16. Manson, J. A. and Sperling, L. H. (1976). Polymer Blends and Composites, Heyden Press, London.
17. Morris, E. R., Rees, D. A., Thom, D. and Welsh, E. J. (1977). *J. Supramol. Struct.* **6**, 259–274.
18. Morris, E. R., Rees, D. A. and Robinson, G. (1980). *J. Mol. Biol.* **138**, 349–362.
19. Parker, T. G. and Dalgleish, D. G. (1977). *Biopolymers* **16**, 2533–2547.
20. Peniche-Covas, C. A. L., Dev, S. B., Gordon, M., Judd, M. and Kajiwara, K. (1974). *Faraday Discuss. Chem. Soc.* **57**, 165–180.

21. Rees, D. A. and Welsh, E. J. (1977). *Angew. Chem. Int. Ed. Engl.* **16**, 214–224.
22. Richardson, R. K. and Ross-Murphy, S. B. (1981). *Int. J. Biol. Macromol.* **3**, 315–322.
23. Richardson, R. K. and Ross-Murphy, S. B. (1981). *Br. Polym. J.* **13**, 11–16.
24. Samejima, K., Ishioroshi, M. and Yasui, T. (1981). *J. Fd Sci.* **46**, 1412–1418.
25. Smoluchowski, M. V. (1917). *Z. Physik. Chem.* **92**, 129–168.
26. Stockmayer, W. H. (1943). *J. Chem. Phys.* **11**, 45–55.
27. Tombs, M. P. (1970). Proteins as Human Foods, ed. R. A. Lawrie, pp. 126–138, Butterworths, London.

Note added in proof: Recent kinetic simulations by the author suggest that equation 14 should also fit modulus-concentration data for irreversibly gelling systems, provided that modulus values are compared at the same time after cross-linking has commenced. K now depends on the forward rate constant and the time of reaction and ceases to be an equilibrium constant.

3 General Principles of Crystallization

J. GARSIDE

Department of Chemical Engineering, UMIST

1 INTRODUCTION

Crystallization can take place from the vapour, melt or solution phase and this chapter will discuss the general theories which describe the nucleation and growth characteristics of crystallization processes. In general, special theories have not been developed to treat these different phase changes but rather general theories have usually been adapted and perhaps modified to apply to specific cases. For example the complicating effects of the solvent through solute/solvent interactions and solvation effects are usually ignored in the description of solution growth and complicated multi-ionic systems are usually treated as "one-unit" systems.

The first two sections will treat the kinetic events of nucleation and growth respectively. The third section will provide an introduction to the mathematics of modelling crystal size distributions in crystallization processes.

FOOD STRUCTURE AND BEHAVIOUR
ISBN 0-12-104230-8

2 NUCLEATION

Crystallization cannot take place until the phase is "supersaturated" or "undercooled". This condition alone, however, is not sufficient for crystallization to occur; new crystallization centres must first exist in the solution. This formation of one phase in another under conditions where a free energy barrier exists is the process of nucleation.

There are various types of nucleation process and these are categorized in Fig. 1. Nucleation which occurs in clean solutions, uncatalysed by the presence of foreign particles or interfaces, is known as primary, homogeneous nucleation. The presence of foreign surfaces catalyses the nucleation process and gives rise to primary, heterogeneous nucleation which occurs at lower levels of supersaturation than homogeneous nucleation. A third nucleation mechanism, secondary nucleation, occurs only because of the prior presence of crystals of the material being crystallized. It has been recognized as a mechanism in its own right particularly as a result of work in the field of industrial bulk crystallization.

The ice/water system provides a vivid example of these processes. Carefully purified water, distilled and filtered can be cooled to below $-30\,^{\circ}\mathrm{C}$ comparatively easily before ice forms by primary, homogeneous nucleation. Tap water can only be cooled to about $-6\,^{\circ}\mathrm{C}$ before primary heterogeneous nucleation occurs. On the other hand a continuous crystallizer operating with a retained bed of ice can be operated satisfactorily at -2 or $-3\,^{\circ}\mathrm{C}$, nucleation taking place continuously through a secondary mechanism.

A Primary homogeneous nucleation

The classical approach to nucleation[2] envisages homogeneous nucleation as occurring by a series of bimolecular reactions between solute species (molecules or ions) giving rise to a distribution of clusters or embryos in the

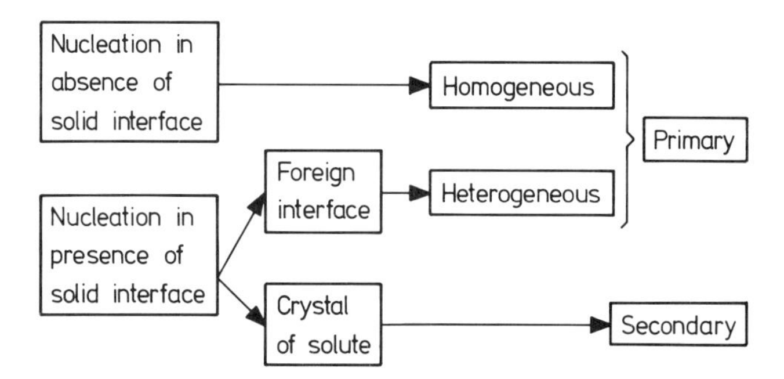

Fig. 1 Classification of nucleation processes

solution, the distribution depending on, among other parameters, temperature and supersaturation. The Gibbs free energy of the embryo ΔG_r results from both a surface and a volume contribution. It is hence size dependent and for a spherical nucleus of radius r is given by

$$\Delta G_r = 4\pi r^2 \gamma - \frac{4\pi r^3 \Delta G_v}{3V_m} \tag{1}$$

where γ is the surface free energy per unit surface area, ΔG_v the molar free energy change associated with the fluid-solid phase change and V_m is the molar volume. In deriving equation (1), it is assumed that the macroscopic surface free energy can be used for clusters containing a comparatively small number of atoms or molecules.

The form of equation (1) is shown in Fig. 2. The overall free energy goes through a maximum at some critical size r^* which can be shown to be given by

$$r^* = \frac{2\gamma V_m}{\Delta G_v} \tag{2}$$

All the embryos will tend to change their size in such a way as to decrease their overall free energy. For those smaller than the critical size this is achieved by dissolving whereas clusters larger than r^* will continue to grow. The critical cluster size therefore corresponds to the smallest cluster that can decrease its free energy by growing and so is the minimum size of a stable nucleus.

The rate at which new nuclei are formed is a problem of kinetics and is determined by the rate at which nuclei surmount the maximum in the free energy curve (Fig. 2), this being the rate limiting step in the nucleation

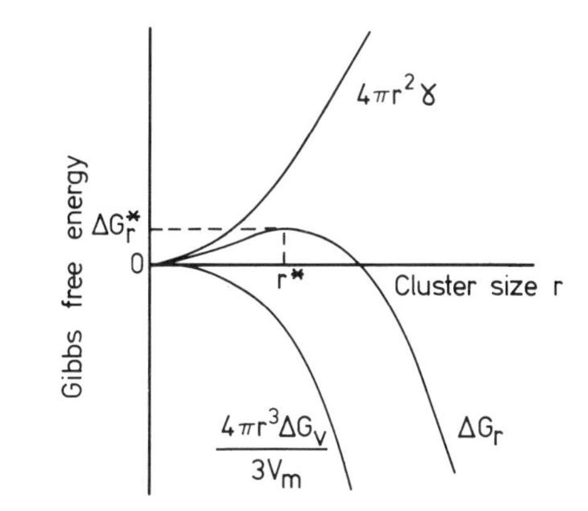

Fig. 2 Free energy of an atomic cluster as a function of cluster radius

process. Volmer[18] wrote the nucleation rate J (number of nuclei per unit time per unit volume) as proportional to $\exp(-\Delta G^*/kT)$ where ΔG^* is the free energy change associated with the formation of the critical nucleus and so obtained

$$J = k_n \exp\left(-\frac{16\pi v^2\gamma^3}{3k^3T^3(\ln S)^2}\right) \tag{3}$$

For crystallization from solution S is the supersaturation ratio, c/c_{eq}, with c_{eq} the equilibrium concentration, and v is the molecular volume.

Attempts have been made to put the above theory on a sounder theoretical basis and a comprehensive review of much of this work is given by Zettlemoyer.[21] Most effort has been devoted to making better estimates of the pre-exponential factor, k_n. The form of equation (3) is however dominated by the exponential term at low supersaturations and so the essential features of the equation remain unchanged no matter what approach is adopted.

Figure 3 illustrates the relation between nucleation rate and supersaturation as predicted by equation (3). Below a critical value of supersaturation, S^*, the nucleation rate is extremely small; once this supersaturation is reached, however, the nucleation rate increases very rapidly. On a plot such as this the values on the ordinate are not important as very small changes in S can produce orders of magnitude changes in J. The supersaturated region thus is divided essentially into two regions; at low supersaturations, the nucleation rate is effectively zero and the solution is "metastable"—crystals will grow but nucleation is negligible. At high supersaturations, rapid nucleation takes place and the solution is "labile". This is frequently illustrated in a "solubility–supersolubility diagram" such as given in Fig. 4.

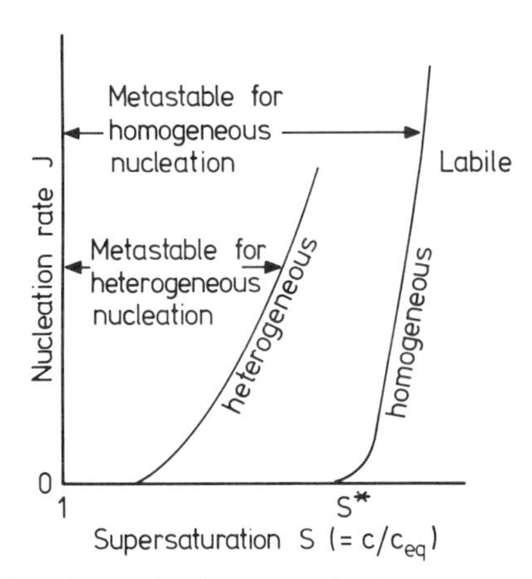

Fig. 3 Form of primary nucleation rate equation

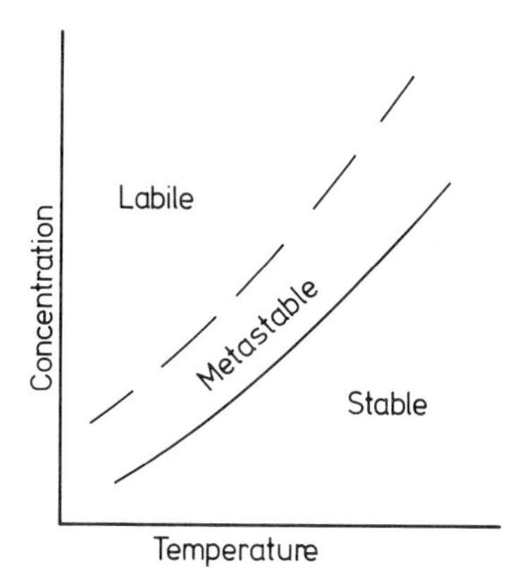

Fig. 4 Solubility–supersolubility diagram

B Primary heterogeneous nucleation

Most primary nucleation that occurs in practice is likely to be heterogeneous, i.e. induced by foreign particles. This requires significantly lower supersaturations than homogeneous nucleation and so the width of the metastable zone is usually greatly reduced.

Nucleation on a foreign particle will, in general, require a lower free energy change than that necessary to induce nucleation in the bulk solution. The exact mechanism by which this occurs is not clear but it is likely to be the result of a local ordering process brought about by the interactions across the interface. Consequently the free energy of formation of a critical nucleus is reduced by a factor ϕ $(0 < \phi < 1)$ and so:

$$\Delta G^*_{r(het)} = \phi\, \Delta G^*_r$$

ϕ has frequently been related to the contact angle θ between the crystal and the foreign substrate as shown in Fig. 5 where a simple force balance gives $\cos\theta = (\gamma_{SL} - \gamma_{CS})/\gamma_{CL}$. ϕ is then given by[19]

$$\phi = (2 + \cos\theta)(1 - \cos\theta)^2/4 \qquad (4)$$

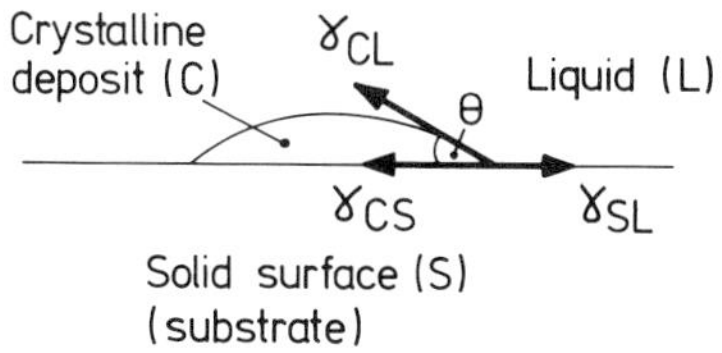

Crystalline deposit (C)
γ_{CL}
Liquid (L)
θ
γ_{CS}
γ_{SL}
Solid surface (S) (substrate)

Fig. 5 Cap-shaped nucleus model of heterogeneous nucleation. The interfacial surface energies and the contact angle θ are illustrated

Turnbull and Vonnegut[17] related the catalytic power of a substrate to the lattice mismatch between it and the growing crystal surface and it is clear that such compatability aids nucleation. The heterogeneous nucleation rate equation appears to be of similar form to that of equation (3) but the catalytic power of the heteronuclei reduces the critical supersaturation at which a rapid increase in nucleation occurs and which defines the limit of the metastable zone as shown in Fig. 3.

There is some evidence that the size of a potential heteronucleus affects its nucleating capabilities and Walton[20] has suggested that the limited amount of data available indicates the range 0.1–1 μm as being the most effective.

C Induction times

Perhaps the most common method of investigating primary nucleation phenomena in solution systems is by the measurement of induction times. A supersaturated solution is prepared, usually by mixing two reactant solutions, and after some time (t_{ind}) crystals are detected. The time that elapses before the detection of crystals depends both on the time taken to form stable nuclei and on the time for their subsequent growth to observable size. It has usually been assumed that the induction time can be used as a measure of the nucleation rate and so, noting that $t_{ind} \propto J^{-1}$, equation (3) can be rewritten as

$$t_{ind} \propto \exp\left[\frac{A\gamma^3}{T^3(\ln S)^2}\right] \tag{5}$$

or

$$\log t_{ind} \propto \frac{\gamma^3}{T^3(\ln S)^2} \tag{5a}$$

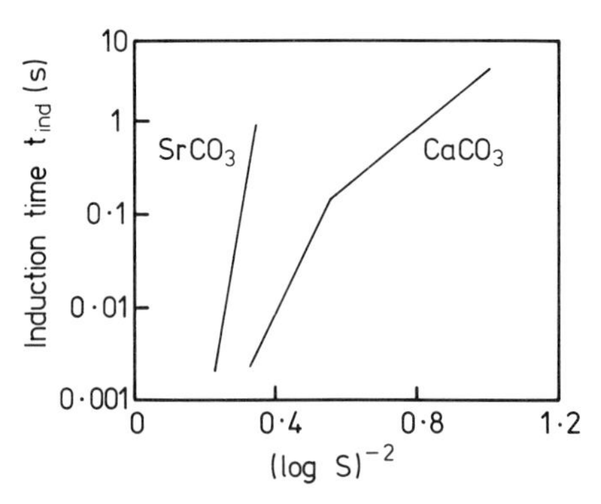

Fig. 6 Induction time as a function of supersaturation for precipitation of $SrCO_3$ and $CaCO_3$ (from ref 14)

Figure 6 is an example of such a plot, the slope enabling the interfacial surface energy γ to be determined.

D Secondary nucleation

Recent reviews of secondary nucleation (e.g. see refs 3 and 5) have demonstrated that it can result from several different mechanisms. For example, under certain conditions fluid shear forces are sufficient to produce secondary nuclei from an existing crystal surface.[10] Needle breeding sometimes arises at high supersaturations where dendrites may develop or needles grow from existing crystals. If these crystals break, new centres for crystal growth are formed. Contact nucleation arises from contacts between two growing crystals or between a crystal and some other solid surface.[5]

Most secondary nucleation studies have been performed in agitated systems such as are usually found in industrial crystallizers and under such conditions contact nucleation is by far the most important. Examples include the crystallization of ice in sugar solutions and fruit juices.[15] The mechanisms by which these various processes take place are not clear. In most cases hydrodynamic interactions between the solution and crystal surface are important, as is the nature of the crystal surface itself. Secondary nucleation rates depend on the supersaturation and Strickland-Constable[16] has suggested that this arises since the initial size distribution of potential secondary nuclei is independent of supersaturation but that the particles are produced in the range of the critical size. An effect of supersaturation is then noted since variations of the critical nucleus size with supersaturation allow different proportions of the original nuclei to survive and grow to populate the observed size range. This has since become known as the "survival theory".

Recent studies have shown that contact nucleation can result in secondary "nuclei" being produced at sizes as large as 30 μm with substantial numbers greater than 5 μm, these sizes being very much larger than the critical nucleus.[6]

3 GROWTH

A The crystal solution interface

The mechanism by which a crystal surface grows is determined by the nature of the crystal/solution interface. When a growth unit reaches the crystal surface it will be tightly bound into the lattice if it attaches to a kink site (see Fig. 7), otherwise it is likely to return to the fluid. The probability of reaching a kink site depends on the concentration of kinks on the surface and so the mechanism and rate of growth depend on the structure of the crystal surface.

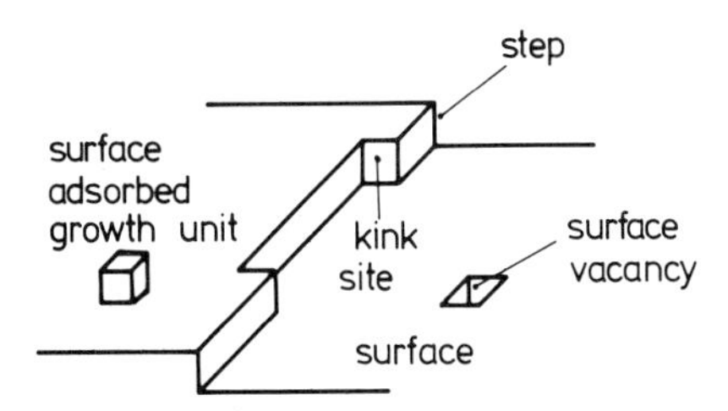

Fig. 7 Schematic view of crystal surface

Qualitatively one can imagine that if the surface is rough on the molecular scale, then many kink sites will be present and "continuous" growth will occur since no thermodynamic barriers to the growth process exist; each growth unit in a supersaturated state that reaches the surface will be incorporated into the crystal. As the surface becomes smoother on the molecular scale, growth becomes more difficult; surface diffusion of growth units must be taken into account and the exact morphology of the surface defines the growth mechanism. Under these circumstances "layer" growth models are appropriate.

The nature of the crystal surface can be specified in terms of the values of two parameters, α and β. α is a surface energy parameter (see, for example [9]) given by

$$\alpha = \frac{4\varepsilon}{kT} \tag{6}$$

where ε is the energy gain on the formation of a solid-fluid bond:

$$\varepsilon = \tfrac{1}{2}(\phi_{ss} + \phi_{ff}) - \phi_{sf} \tag{7}$$

the ϕs being bond energies.

The driving force for crystallization is the difference in chemical potential, $\Delta\mu$, between a growth unit in solution and in the crystal. The dimensionless parameter β is defined by

$$\beta = \frac{\Delta\mu}{kT} \tag{8}$$

which for solution growth can be written

$$\beta \simeq \ln(c/c_{eq}) = \ln(1 + \sigma) \simeq \sigma \tag{9}$$

σ being the supersaturation given by $\sigma = (c - c_{eq})/c_{eq}$. The approximation for β in equation (9) is valid at low supersaturations such as are generally found in solution growth.

In general at low values of α the surface is rough and "continuous" growth will occur. As α increases the surface becomes smoother and other growth mechanisms are then necessary in order for growth to take place. High values of both β (i.e. supersaturation) and temperature favour roughening of the surface. Calculated values of α at which such transitions occur depend on the theoretical treatment adopted but if α is less than about 2.5 or 3 continuous growth will occur. Between these values and 4 or 5, two-dimensional nucleation growth mechanisms are likely while at still higher values of α the presence of steps on the crystal surface is necessary for growth to take place.

B Crystal growth models

CONTINUOUS GROWTH

If all growth units arriving at the surface find a site for incorporation then for small supersaturations the growth rate v is related linearly to the supersaturation:

$$v = C\sigma \tag{10}$$

If the roughness is great enough, all anisotropy may vanish and under these circumstances the crystal will no longer form flat faces or "facets".

SURFACE NUCLEATION MODELS

In order for a two-dimensional nucleus to form on a crystal surface, sufficient growth units must cluster together to form a stable critical nucleus. Once this has been achieved, the edge of this nucleus provides an attachment point for other growth units fed by surface diffusion and lateral growth can proceed on an otherwise flat surface. A number of different nucleation theories of growth have been proposed differing mainly in the relative time scales for the nucleation as compared to the spreading process. Ohara and Reid[11] discuss three such models which all result in the growth rate/supersaturation relation being of the form

$$v = A\sigma^p \exp(-B/\sigma) \tag{11}$$

where p has values between $-\frac{3}{2}$ and $\frac{5}{6}$ and A and B are constants.

The characteristic of all these models is that the exponential term dominates at low values of supersaturation (σ) to give near zero growth rates until some critical supersaturation has been reached.

CONTINUOUS STEP MODELS

For crystal surfaces with high α values, the interface is so smooth that a high energy barrier prevents the formation of two-dimensional nuclei at low

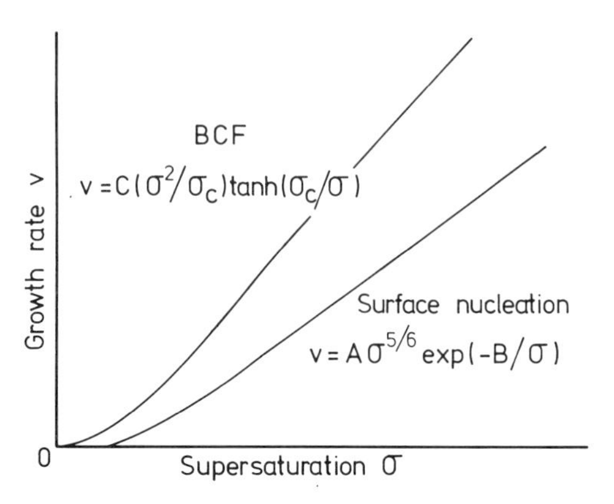

Fig. 8 Form of growth rate equations

supersaturations. In such cases, Frank[4] suggested that the presence of emergent screw dislocations would produce steps and hence kink sites on the surface. Burton, Cabrera and Frank (BCF)[1] subsequently developed a step model of crystal growth in which essentially flat crystal surfaces grow by the addition of growth units to kink sites in an infinite sequence of equidistant steps. The steps are again fed by surface diffusion. The BCF theory then predicts the following relation between growth rate and supersaturation:

$$v = C(\sigma^2/\sigma_c)\tanh(\sigma_c/\sigma) \tag{12}$$

σ_c being a constant specific to the system. The form of both equations (11) and (12) is shown in Fig. 8.

For correlating purposes the growth rate/supersaturation relation for all these models can be conveniently written in a power law form:

$$v = K_G(c - c_{eq})^g \tag{13}$$

C Effect of bulk diffusion

Before growth units reach the crystal surface, they must diffuse from the bulk solution through the concentration boundary layer to the crystal/solution interface. In general, therefore, the supersaturation available at the crystal surface to "drive" the growth process will be less than that based on measurement of the bulk solution concentration. The bulk diffusive process is linear in the concentration difference between the bulk and interface concentrations and is also sensitive to the fluid mechanics of the crystal/solution flow system. Growth rates measured under conditions of bulk diffusive control clearly do not represent the interfacial kinetics. Figure 9 is an example

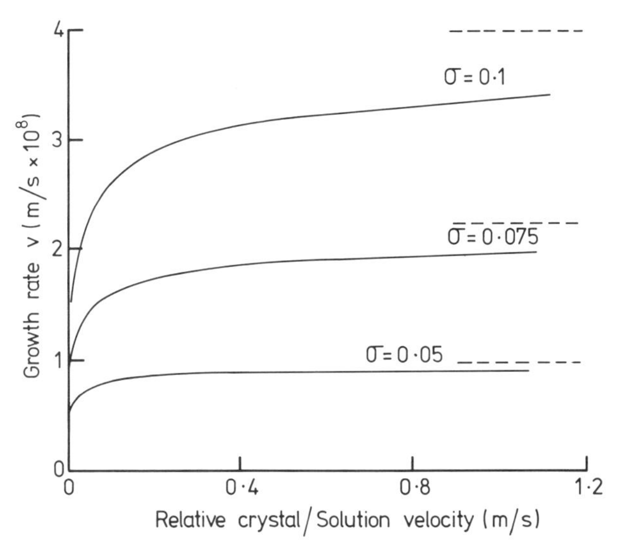

Fig. 9 Effect of solution velocity and hence bulk diffusion, on crystal growth rate

of measurements made under such circumstances. As the relative velocity between crystal and solution is increased the crystal growth rate approaches the surface integration rate where the growth is unimpeded by the bulk diffusion process.

4 CRYSTAL SIZE DISTRIBUTION

The crystal size distribution produced in a crystallization process is of central importance in determining the suitability of the product for subsequent processing and for customer use. Because of this importance, the calculation and control of size distribution has been a central theme of much recent crystallization research. Early work in this field has been discussed by Randolph and Larson.[13]

The general approach is based on solution of population or number balance equations. To provide an introduction to the method, the simple case of a continuous mixed suspension mixed product removal (MSMPR) crystallizer operating at steady state conditions can be considered. In such a crystallizer, the slurry (or suspension) is well mixed and there are no spatial variations of composition or crystal size distribution. Further, the product removed from the vessel is identical in every way to the vessel contents. All crystals are taken to be of the same shape, their size being represented by a characteristic length L. Crystal growth rate G is given by the time variation of L

i.e.
$$G = \mathrm{d}L/\mathrm{d}t \tag{14}$$

If there are no crystals in the feed stream, there is negligible crystal breakage and agglomeration and the crystal growth rate does not vary with crystal size, then the size distribution is given by

$$n = (B/G)\exp(-L/G\tau) \tag{15}$$

where n is the number density and τ is the mean residence time of both fluid and solid phases in the crystallizer and is determined from the ratio of crystallizer slurry volume to volumetric flow rate through the crystallizer. B is the nucleation rate (number of crystals produced per unit time per unit slurry volume).

The assumptions made in deriving equation (15) can often be closely matched in laboratory crystallizers and even in a number of industrial units. Its most valuable use, however, is as a laboratory vehicle for the determination of growth and nucleation kinetics via measured steady state size distributions. A typical experimental size distribution is shown in Fig. 10, the plot of log n vs L being suggested by equation (15). Since the mean residence time is an independently measured variable, the crystal growth rate G can be determined from the slope of this plot and then the nucleation rate, B, evaluated from the intercept at $L = 0$. An evaluation of published kinetic data deduced in this way has been made by Garside and Shah.[7]

The number, length, area and mass distributions can all be calculated from

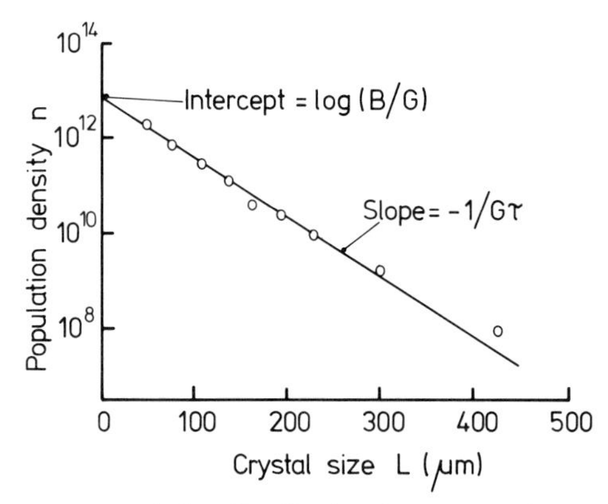

Fig. 10 Steady-state crystal size distribution from a continuous MSMPR laboratory crystallizer: crystallization of α-lactose from pure lactose solution at 30 °C (from ref 8)

equation (15). For example, the differential mass fraction distribution is given in dimensionless form by

$$m(x) = \tfrac{1}{6}x^3 \exp(-x) \tag{16}$$

while the cumulative mass fraction distribution is

$$M(x) = (1 + x + \tfrac{1}{2}x^2 + \tfrac{1}{6}x^3) \exp(-x) \tag{17}$$

where $x = L/G\tau$, the dimensionless crystal size.

The mode of this mass distribution occurs at $x = 3$ while the mass median size can be shown from equation (17) to be at $x = 3.67$. These two mass distributions are shown in Fig. 11. As a measure of the spread of this distribution, it can be shown that the coefficient of variation, equal to the ratio of the standard deviation to the mean, is 50%.

For the continuous MSMPR crystallizer, the shape of the size distribution described by equations (15)–(17) cannot be changed although the position of the distribution on the size axis can be changed by varying the product $G\tau$. Thus, higher growth rates and longer residence times will both increase the crystal size. In practice, such changes are not easy to achieve. For example, a longer residence time will result in a lower level of supersaturation in the crystallizer since the solution now has longer to deposit existing supersaturation on the available crystal surface area. The growth rate will therefore be reduced and the increase in τ is largely offset by this decrease in G. Changes in residence time are thus inefficient in producing changes in crystal size.

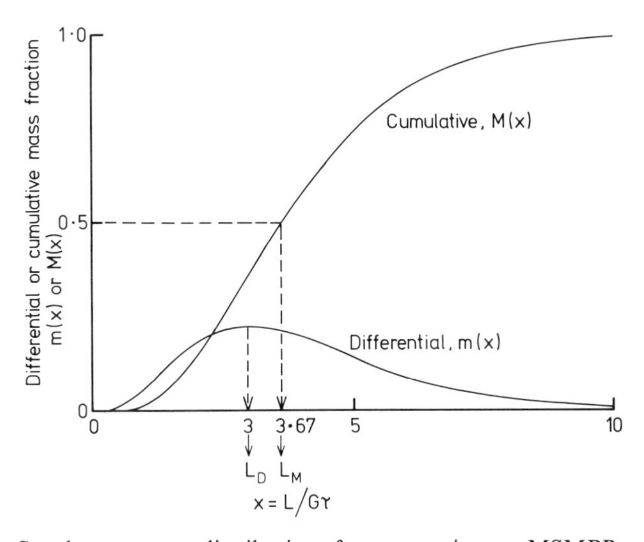

Fig. 11 Steady state mass distributions from a continuous MSMPR crystallizer

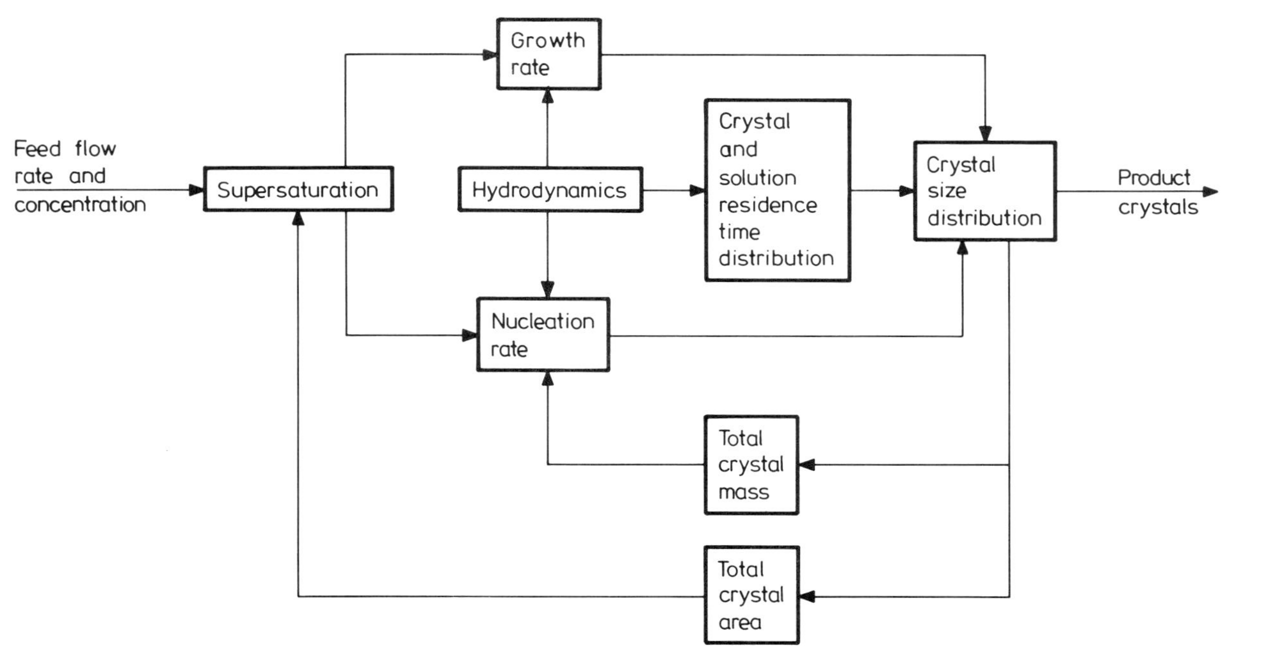

Fig. 12 Crystal size distribution interactions within a crystallizer

The feedback relationships that govern these processes are illustrated in Fig. 12. They have been the subject of extensive theoretical study including much work on the dynamic behaviour of continuous crystallizers and the characteristics of batch systems. The review by Randolph[12] is a good introduction to this aspect of crystallizer operation.

References

1. Burton, W. K., Cabrera, N. and Frank, F. C. (1951). *Phil. Trans. Roy. Soc. Lond.* **243**, 299.
2. Dunning, W. J. (1969). Nucleation, ed. A. C. Zettlmoyer, Dekker, New York.
3. Estrin, J. (1976). Preparation and Properties of Solid State Materials, ed. W. R. Wilcox, vol. 2, pp. 1–42. Dekker, New York.
4. Frank, F. C. (1949). *Disc. Farad. Soc.* **5**, 48.
5. Garside, J. and Davey, R. J. (1980). *Chem. Eng. Commun.* **4**, 393.
6. Garside, J. and Larson, M. A. (1978). *J. Crystal Growth* **43**, 694.
7. Garside, J. and Shah, M. B. (1980). *Ind. Engng. Chem. Proc. Des. Dev.* **19**, 509.
8. Griffiths, R. C. and Merson, R. L. (1982). *A.I.Ch.E. Symposium Series No.* 218 **78**, 118.
9. Jackson, K. A. (1958). Liquid Metals and Solidification, American Society for Metals, Cleveland, p. 174.
10. Jagannathan, R., Sung, C. V., Youngquist, G. R. and Estrin, J. (1980). *A.I.Ch.E. Symposium Series No. 193* **76**, 90.
11. Ohara, M. and Reid, R. C. (1973). Modelling Crystal Growth Rates from Solution, Prentice-Hall, Inc., Englewood Cliffs, N.J.
12. Randolph, A. D. (1980). *A.I.Ch.E. Symposium Series No. 193* **76**, 1.
13. Randolph, A. D. and Larson, M. A. (1971). Theory of Particulate Processes, Academic Press, New York.
14. Sohnel, O. and Mullin, J. W. (1978). *J. Crystal Growth* **44**, 377.
15. Stocking, J. H. and King, C. J. (1976). *A.I.Ch.E.J.* **22**, 131.
16. Strickland-Constable, R. F. (1972). *A.I.Ch.E. Symposium Series No. 121* **68**, 1.
17. Turnbull, D. and Vonnegut, B. (1952). *Ind. Eng. Chem.* **44**, 1292.
18. Volmer, M. (1939). Kinetik der Phasenbildung, Dresden and Leipzig: Steinkopf.
19. Walton, A. G. (1967). The Formation and Properties of Precipitates, Interscience, New York.
20. Walton, A. G. (1969). Nucleation, ed. A. C. Zettlemoyer, pp. 225–307. Dekker, New York.
21. Zettlemoyer, A. C. (ed), (1969). Nucleation, Dekker, New York.

4 Ice Crystallization and its Control in Frozen-Food Systems

J. M. V. BLANSHARD and F. FRANKS

University of Nottingham and PAFRA Ltd, Cambridge

1 INTRODUCTION

Our aim is to explore briefly the importance of freezing in the preservation and processing of foods, and indeed biological systems generally, and yet at the same time discuss modifications or even alternative low temperature strategies in foodstuffs. Since all three issues are important scientifically and technologically, then we must consider the physics and physical chemistry of the constituent processes.[1,14,15,27]

To the food processor, water, in terms of the physical form at ambient temperatures is relatively harmless. However, it *is* the medium in which the various biochemical processes proceed *post mortem* or after harvest which, for many foodstuffs, lead to a chemical deterioration of product quality. In general the processes obey an Arrhenius type relation to temperature and therefore are faster at higher temperatures and slower at lower temperatures.

FOOD STRUCTURE AND BEHAVIOUR
ISBN 0-12-104230-8

An obvious recipe for reducing the rates of such reactions is to lower the temperature down to but *not* below the freezing temperature; chill preservation is the application of this principle. The usual industrial minimum is −2 °C.

In contrast, from a chemical point of view, it might be thought that the freezing event and conversion of water to ice is innocuous since the effects of both reduced temperature and ice separation lead beneficially to the further retardation of chemical processes. In practice, various workers have reported a frequently undesirable enhancement of reaction rates in *partially* frozen systems[11,19] due probably to freeze-concentration while, in sharp contrast to the situation above 0 °C, the change in physical form is devastatingly destructive with gross mechanical disruption of structures through ice crystallization.

2 THE PHYSICAL FORMS OF ICE

It is important, however, at this stage to note the various crystallographic forms of ice (eight polymorphic forms) all of which, excluding ice I, are only

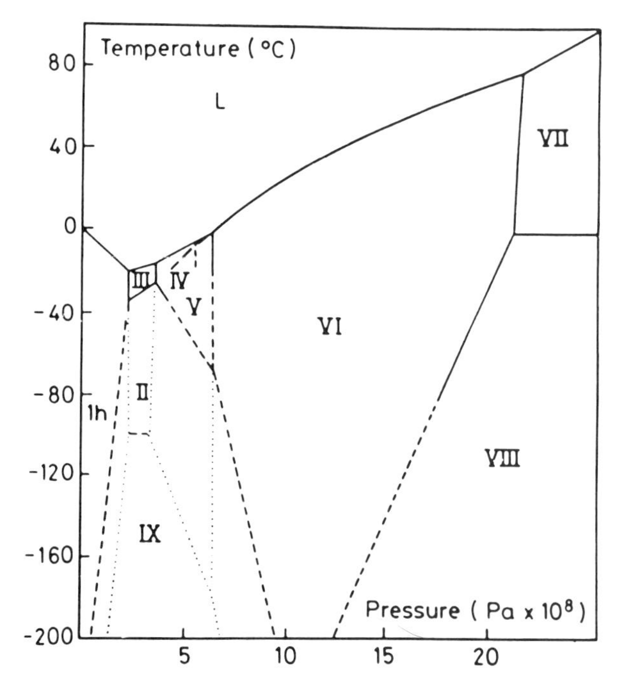

Fig. 1 Solid-liquid phase diagram of H_2O. Broken lines represent approximate and dotted lines estimated or extrapolated phase boundaries. Numbers I–VIII are the different polymorphic forms

stable at high pressure (Fig. 1). In addition to these crystalline forms, solid water in the amorphous or glassy state has also been reported. Although this is not readily formed in the pure state by very-rapid freezing, it may be prepared in a high degree of purity by condensation from vapour at temperatures less than −173 °C.

The glassy state is itself still a matter of research and debate but it is characterized by greatly reduced molecular mobility, one related and widely quoted criterion being that it is a disordered system with a viscosity not less than 10^{14} Pa/s. The physical significance of such a value is better appreciated if we note that the time taken for a water molecule to diffuse a distance of 1 cm is of the order of 300 years. Such a system would presumably minimize any chemically deteriorative processes. The potential value of the glassy state is more readily seen by studying a state diagram as in Fig. 2 for the water-sucrose system. The glassy region is shaded, and though with pure water this occurs only at temperatures below −133 °C, with pure sucrose, T_g is 52 °C. It is also evident that there is a significant concentration regime where the glassy state will be stable at temperatures above −20 °C, the normal domestic freezer temperature. It must be emphasized, however, that the glassy state is only readily attained by a limited number of systems, namely

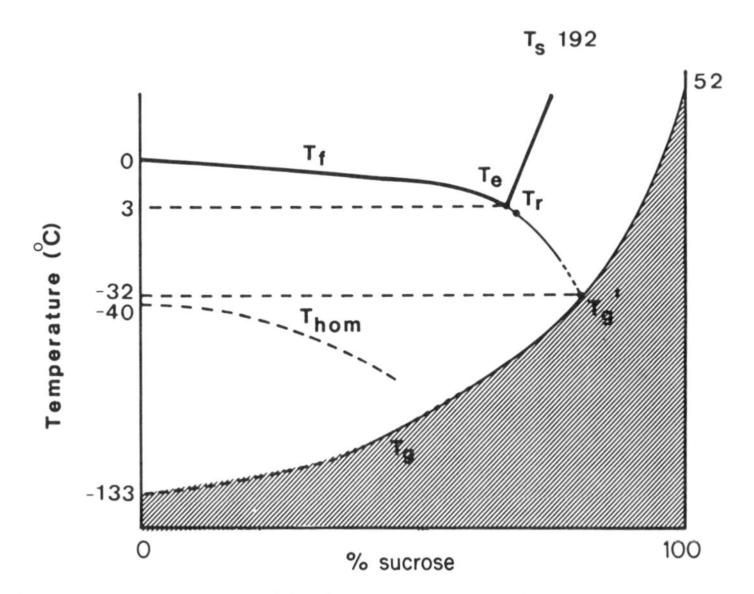

Fig. 2 Water-sucrose composition/temperature state diagram. Bold lines represent equilibrium phase coexistence curves. Other lines denote processes under kinetic control. Broken lines indicate probable undetermined relationships. The shaded region is the glassy state

those where the solute does not easily crystallize from solutions but rather forms highly supersaturated solutions on concentration.

3 ICE CRYSTALLIZATION

Some consideration has already been given in the previous chapter to the process of crystallization which has been dissected into the constituent events of nucleation (both homogeneous and heterogeneous), crystal growth and recrystallization or so-called "Ostwald ripening". These processes all play a role in ice crystallization.

4 HOMOGENEOUS NUCLEATION

Homogeneous nucleation is that process whereby clusters of molecules arise spontaneously by random density fluctuations. In thermodynamic terms the equilibrium freezing point ($T_E = 273.16$ K) indicates that temperature at which the molar Gibbs free energies (chemical potentials) of the liquid and ice states are the same. The formation of a nucleus leads to an unfavourable, positive surface free energy but a negative volume free energy change. The sum of these two components and the relative contributions of these to the overall free energy as the radius increases is the reason for the existence of a critical nucleus size at any given temperature. Since r^* falls as the temperature is reduced, it is not surprising that the probability of nucleation approaches unity at some low temperature, actually -41 °C. This is the so-called homogeneous nucleation temperature (T_h). Pure liquid water can therefore not exist at atmospheric pressure at temperatures below T_h.

We may usefully employ the classical theory of nucleation, despite a number of dubious assumptions, to calculate the rate of nucleation. It may be shown[16,27] that the rate of formation of nuclei per unit volume, $\boldsymbol{J}$ ($\mathrm{m^{-3}s^{-1}}$) is a function of temperature of the following form:

$$\boldsymbol{J} \simeq K_1(\sigma T)^{1/2}\phi^2 \exp(-\Delta G^{\ddagger}/kT)\exp(-K_2\sigma^3/\Delta T^2 T^3)$$

where σ is the macroscopic surface energy of the ice-solution interface, ϕ is the volume fraction of water in the solution, $\Delta G^{\ddagger}$ is the free energy of activation of diffusion of the slowest moving components of the solution and K_1 and K_2 are obtained from fundamental constants relating to pure water and have the values:[27]

$$K_1 \simeq 2.40 \times 10^{40}\, \boldsymbol{J}^{-1/2}\, \mathrm{m^2 s^{-1}}$$

$$K_2 \simeq 6.06 \times 10^{16}\, \mathrm{m^2} K^5 \boldsymbol{J}^{-3}$$

Muhr *et al.*,[28] have pointed out that solutes can affect $\boldsymbol{J}$ for a given value of ΔT by:

i altering σ. Fig. 3 curves 1, 2 and 6 show that $\boldsymbol{J}$ is greatly affected by small changes in σ because of the occurrence of σ^3 in the $\boldsymbol{J}(T)$ equation.
ii altering $\Delta G^{\ddagger}$. Solutes increase $\Delta G^{\ddagger}$ at temperatures greater than T_E but by reducing the "structuring" of water they may tend to affect the rise in $\Delta G^{\ddagger}$ observed in undercooled water.
iii reducing T_E. This will reduce $\boldsymbol{J}$ at a given value of ΔT chiefly because $\exp(-K_2\sigma^3/\Delta T^2 T^3)$ gets smaller as T falls (for constant ΔT and σ) but the lower value of T should also lead to a smaller value of $\exp(-\Delta G^{\ddagger}/kT)$. In comparison, the pre-exponential factor of $T^{1/2}$ can be regarded as virtually constant.
iv altering ϕ. Since ϕ occurs only as a pre-exponential factor this effect will not be significant.

It is also, of course, possible to express the volume of the nucleus in terms of the number of molecules if we assume that the clusters are ice-like. Similarly the calculation can be extended to determine the proportion of drops of specified size derived from a given mass of water that actually contain a

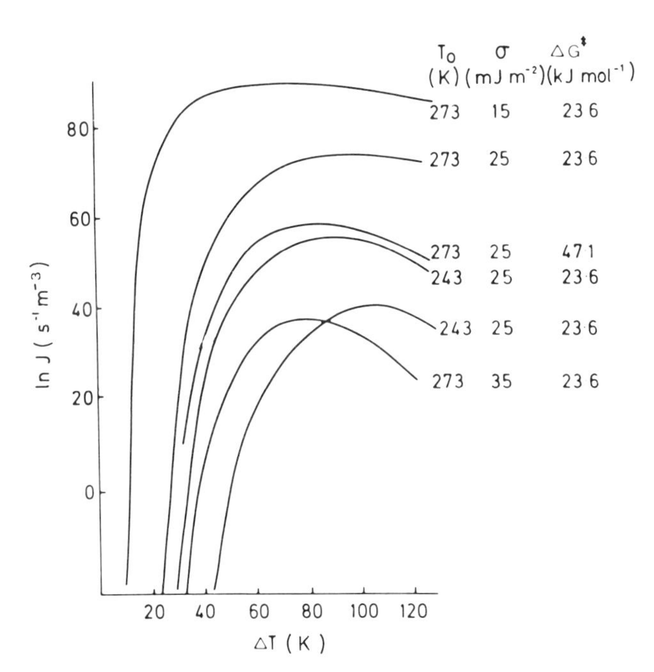

Fig. 3 Rate of homogeneous nucleation ($\boldsymbol{J}$) as a function of undercooling (ΔT) with σ, T_E and $\Delta G^{\ddagger}$ as parameters

Table 1 Distribution of ice nuclei within dispersed water droplets[10]

Temperature (K)	Radius of drop (nm)	r^* (nm)	Droplets g^{-1}	Nuclei g^{-1}	Proportion of droplets containing a nucleus
263	10	4.9	2.4×10^{17}	2×10^{-433}	8.4×10^{-449}
253	10	2.26	2.4×10^{17}	8.2×10^{-75}	3.4×10^{-92}
	5	2.46	1.9×10^{18}	2.5×10^{-93}	1.3×10^{-111}
243	10	1.46	2.4×10^{17}	2.5×10^{-18}	1.1×10^{-35}
	5	1.55	1.9×10^{18}	1.4×10^{-23}	7.2×10^{-42}
233	10	1.08	2.4×10^{17}	6.3	2.6×10^{-17}
	5	1.14	1.9×10^{18}	0.029	1.5×10^{-20}

nucleus and will therefore freeze at a given temperature. These figures are shown in Table 1. These data amply demonstrate that nucleation is extremely sensitive to changes in temperature and that the very large numbers involved in the calculations change rapidly by many orders of magnitude over small ranges of temperature. It is also evident that at −20 °C the proportion of droplets (of diameter 5 or 10 nm) containing an ice nucleus is extremely small.

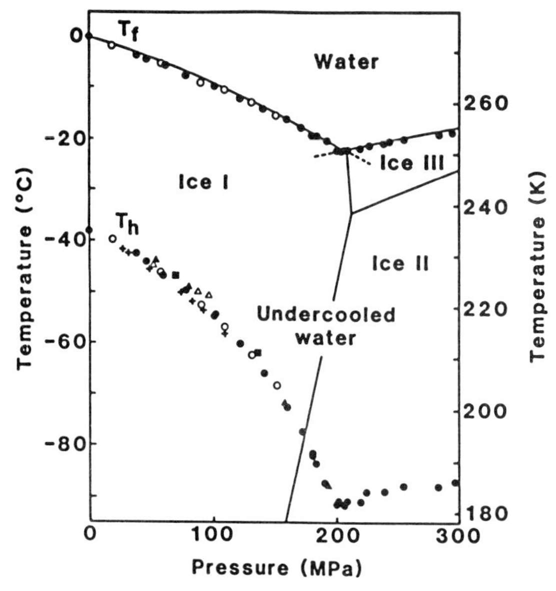

Fig. 4 The effect of pressure on the homogeneous nucleation of ice in undercooled water, according to Angell[1]

The other thermodynamic variable of significance is pressure and this does exert a depressive effect on the homogeneous nucleation temperature. Indeed the effect of pressure on T_h is twice that on T_E (Fig. 4).

5 HETEROGENEOUS NUCLEATION

Pure water without any solid or liquid contaminating substrates is very rare in practice and therefore these other factors need to be borne in mind. We can foresee from our previous discussions that the intrusion of a surface active mote into a system introduced a region of reduced surface-free energy which thereby enhances the probability that a cluster of critical dimensions will form. This so-called process of heterogeneous nucleation can be distinguished from homogeneous nucleation through the use of calorimetric techniques and the elegant emulsification method of Rasmussen and Mackenzie.[30] The process of emulsification is extended to the point where the motes which are capable of inducing heterogeneous nucleation are confined to a negligible proportion of the emulsified droplets.

There is experimental evidence that some materials are particularly active as catalysts of nucleation and conversely others that are not. Certain micro-organisms, insects and plants contain materials which possess high concentrations of very efficient ice nucleators.[22,24] Most biological cells contain structures that are able to catalyse the nucleation of ice to some degree, e.g. the plasma membrane of human erythrocytes only slightly promotes the freezing of intracellular water which occurs at a temperature of $T_h + 0.5\,^{\circ}\mathrm{C}$ above the T_h of isotonic saline solution. Yeast cells, however, have sites which substantially promote freezing at as high as $T_h + 10\,^{\circ}\mathrm{C}$.

Fletcher[12] has pointed out that we can represent the contribution of various factors to the process of heterogeneous nucleation by the expression

$$\Delta G^* \text{ het} = \Delta G_h^* f(m, R)$$

where R is the radius of the catalytic particle (which is assumed to be spherical) and m is a wetting parameter ($-1 < m < 1$), which describes the relative ease of wetting of the particle by ice and undercooled water respectively. For a maximum catalytic efficiency, $R > 10$ nm and $m \simeq 1$ (i.e. wetting by ice) and for a particle to possess any nucleating properties, R must exceed 1 nm.

Where unicellular structures either naturally or through the presence of additives have $T_{het} \simeq T_h$, it is possible to foresee that such materials may be preserved in the aqueous unfrozen state at e.g. $-20\,^{\circ}\mathrm{C}$, thereby benefitting from the low temperature but without the damaging effects of freeze concentration, ice crystal nucleation and growth.

6 "OSTWALD RIPENING"

A process which is of great importance in the frozen storage of food materials is "Ostwald ripening". The initial freezing process, particularly where this occurs rapidly, will yield a highly dispersed crystal phase with a very high surface:volume ratio. This characteristic inevitably promotes metastability a consequence of which is that maturation occurs with growth of the crystals and a broadening of the size distribution of the crystals. The situation is well exemplified by the results for beef muscle frozen at −40 °C and subsequently stored at −5 °C (Fig. 5).

7 A PRACTICAL EXAMPLE—ICE CREAM

Ice cream is an interesting example to consider since its preparation is not primarily for the sake of preservation (as with many frozen products that thereafter are thawed prior to consumption) but to develop certain textural and organoleptic features which are inherent to the frozen state.

The manufacture of ice-cream involves a complex sequence of events (e.g. mixing, pasteurization, homogenization, cooling, ageing, aeration) which culminate in extrusion at −6 °C, hardening at −35 °C and storage/distribution at −11 °C.[3] These temperature shifts lead to changes in the frozen

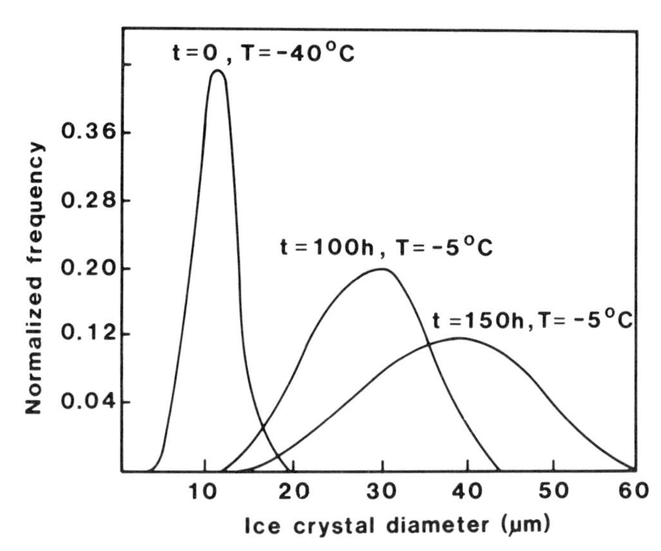

Fig. 5 The effect of frozen storage at −5 °C on the dimensions of ice crystals in beef, with changes, from [6]

water which increases from 50% during extrusion to 95% during hardening and then falls to 72% during storage. During the hardening stage the residual aqueous phase (excluding ice) is a concentrated (86%) sugar glass containing sucrose and lactose; it has a T'_g of −23 °C.

Two phenomena are of particular interest. First, the size of the ice crystals themselves is important and a mean diameter of 40 μm has been shown to be organoleptically acceptable; the distance between crystals is then 6–8 μm. However, different freezing rates, formulations and stabilizers (as we shall see in the next section) and storage temperatures may result in significant modifications. Secondly, the phenomenon known as "grittiness" is due to the crystallization of lactose which is detectable at > −23 °C and becomes rapid at > −17 °C. It is prevented by reducing the milk serum solids.

8 CONTROL OF ICE CRYSTAL GROWTH

In view of the importance of ice-crystals in frozen foodstuffs as a factor in modifying the texture of the materials, it is not surprising that there have been considerable empirical efforts to control ice crystal size and more recently scientific investigations to account for the reported effects of certain additives. It has been reported, for example, that polysaccharides such as carrageenan, alginates and cellulose derivatives and other stabilizers such as gelatin had a beneficial effect in reducing ice-crystal growth in ice cream. In an extensive examination of this problem Muhr[27] found no evidence to support the view that polysaccharide stabilizers affected the amount of ice formed (which is in agreement with physicochemical predictions) or that such stabilizers of themselves enhanced the process of homogeneous or heterogeneous nucleation in concentrated sucrose solutions (except by the introduction of extraneous motes with the stabilizer) in any way that might explain the reported reduction in ice crystal size. Some reduction in the rates of ice crystal growth was noted when stabilizers were present at high sucrose concentrations and a profound change of both crystal type and growth rate was noted where conditions permitted the formation of a gel structure within the system. The observed reduction in growth rate of ice crystals at high sucrose concentrations in the presence of stabilizer is believed to be a consequence of the fact that we have a ternary rather than a binary system, i.e. the sucrose renders stabilizers more effective at bringing about a freezing point decrement and especially so in the high concentrations prevailing in front of an advancing ice interface. Some support for this view comes from the observation that the freezing-point curve for polymeric solutes is generally highly non-linear and although $\partial \Delta T / \partial c$ is almost zero at low c, it becomes quite appreciable at higher values.[18] Indeed an examination of theoretical models of $\ln a_w$ in such systems suggests that this steepening effect of the freezing point curve does not depend so much on the polymer concentration as on the total solute concentration, so that the addition of a low concentration of polymer may reduce the freezing point of a concentrated sucrose solution by much more than it would reduce the freezing point of water.[27]

9 ANTI-FREEZE PEPTIDES

DeVries *et al.*[7,8] were the first to report a thorough study of the glycopeptides found in the body fluids of Antarctic fish species at a concentration of approximately 4% and which had the observed properties of depressing the freezing point of water and in limiting the development of ice crystals. The Antarctic fish peptides (AFPs) have a molecular mass in the range 2600–23 500 daltons and consist of a basic tripeptide unit ala-ala-thr with the disaccharide side-chain galactose-*N*-acetyl-galactosamine attached to each threonyl residue. That this depression of the freezing point of water is not due to a colligative effect is evident from more detailed studies which have shown that the melting temperature reflects the very small depression one might expect from a solute with a mean molecular mass of 10 000 daltons.

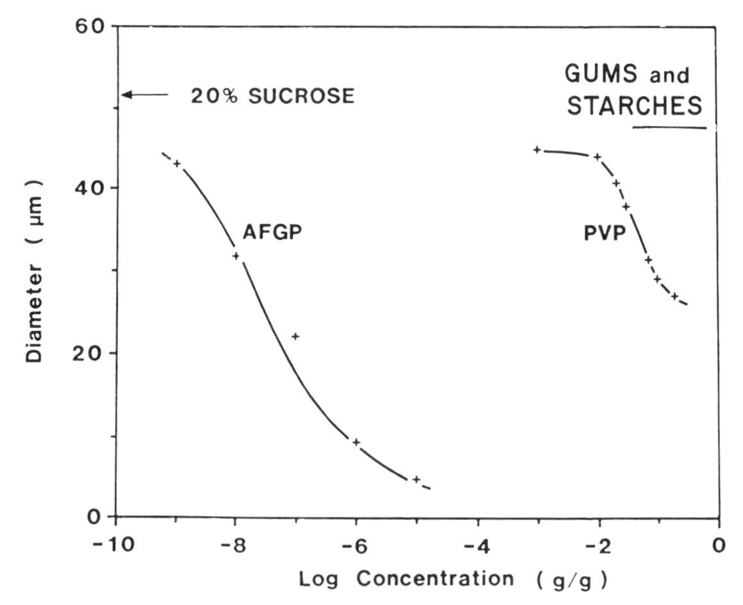

Fig. 6 Size of ice crystals as a function of the concentration of three potential inhibitors of ice crystal growth (AFP, PVP and polysaccharide)[17]

The AFPs also have a marked effect on the morphology of ice crystals grown in their solutions. Several theories have been presented as to how AFPs function. Most of these assume that AFPs are adsorbed at the ice-solution interface in such a way that they inhibit the process of accretion of water molecules to the ice-lattice. Brown *et al.*[5] have presented direct evidence that adsorption of AFPs does occur using a technique based on the second harmonic generation of light at the ice-water interface.

Franks *et al.*[17] have now shown that AFPs are primarily effective, not in preventing ice nucleation, but in limiting ice crystal growth. The relative merits of AFP, polyvinylpyrrolidone and polysaccharide stabilizers in controlling ice crystal growth are shown in Fig. 6 where the mean ice crystal diameter (after 3 h growth at 1.5 °C below the melting point of ice in the solution) as determined by video microscopy is plotted against the concentration of AFP and other polymers in a nucleated 20% sucrose solution.

10 PRACTICAL STRATEGIES FOR CONTROL OF ICE CRYSTALS IN FOOD SYSTEMS

When we review the general theoretical structure and factors responsible for ice crystal nucleation and growth we can propose a series of strategies depending on our ultimate objectives with any food product or biological system. They are itemized below.

i The inhibition of nucleation. Where there is an attendant reduction in temperature, there are the benefits of a minimization of chemical and physical processes without the deleterious effects of freezing and freeze concentration. This approach has been pursued by the Rich Corporation in a series of patents and products where the freezing point has been lowered by the introduction of massive quantities of osmotically active materials, e.g. sugars whose undesirable organoleptic qualities have been minimized by the further addition of astringent agents. A very different approach has utilized the droplet emulsion principle.[26]

ii The control of nucleation. Since ice nucleation and growth are temperature-dependent rate processes with optima at different temperatures, then the relative rates of nucleation and growth of ice crystals at different temperatures may be exploited by the appropriate manipulation of the rates of heat transfer[9] and the physicochemical milieu of the system.[31] In some cases, small ice crystals may be desirable (the consequence of extensive nucleation), in other instances large ice crystals are the ideal and are produced through limitation of undercooling and nucleation.

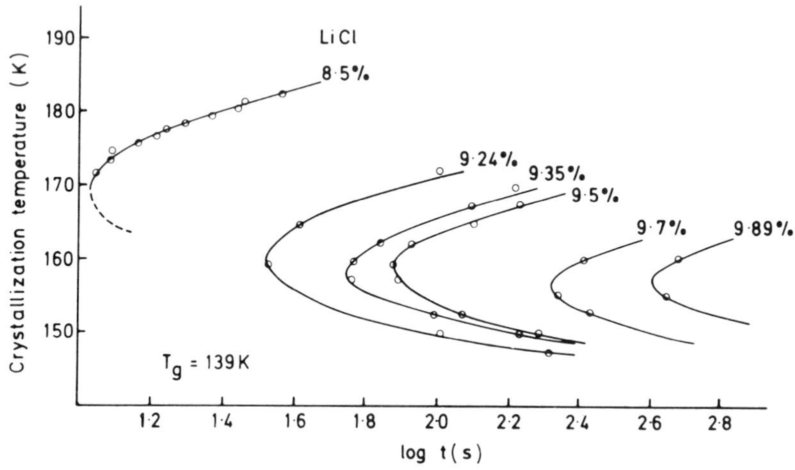

Fig. 7 Time-temperature-transformation (TTT) curves of ice crystallization in concentrated LiCl solutions[25]

The overall effect is conveniently expressed by time-temperature-transformation (TTT) curves which relate the time taken to crystallize a given fraction of the undercooled liquid at a given temperature. Ice crystallization in a concentrated LiCl solution is shown in Fig. 7.

iii The control of ice crystal growth. In this instance, crystals are desired but of the right size. As with AFPs one approach may be to inhibit partially the accretion of water molecules at the ice interface. Alternatively, the presence and accumulation of both micro- and macro-molecular additives may modify the diffusion/colligative properties at the ice crystal water interface and thereby limit extensive ice crystal growth or recrystallization. Undoubtedly in some food systems the macromolecules (proteins and/or polysaccharides) do form a gel network at low if not ambient temperatures which, as has been demonstrated, severely curtails crystal growth and modifies the crystal habit.[29]

iv Exploitation of the glassy state. This area has been the subject of a number of investigations, some of which have emerged from research primarily directed at cryopreservation of tissues.[13] Such workers, though recognizing formidable outstanding problems, nevertheless propose methods whereby vitrification is an intriguing but distinct possibility for the prolonged preservation of complex biological systems in general. In a rather different vein, Boutron[4] has examined from both experimental and theoretical points the notion that, depending on the rate of cooling in a system, so the fraction of water in the ratio of vitrified to crystalline state will vary in a systematic fashion. There would, therefore, be an obvious and exploitable correlation between the kinetics of ice crystallization and the degree of cell damage (for a cellular system) depending on whether the ice crystallizes incompletely inside or outside the cells. An even more interesting development has been the conclusions of Levine and Slade[23] who have demonstrated that T_g' (see Fig. 2) is a function of molecular mass and have discussed how the use of appropriate raw materials in a formulated food product permit one to manipulate T_g' and thereby promote product stability, in this instance minimizing or controlling ice-crystal growth and/or maturation.

11 THE SINTERING OF ICE CRYSTALS

A somewhat different problem, which is nevertheless of considerable importance to the purveyor of frozen foods is that under certain conditions, frozen particulate food products that have been prepared with the utmost care as discrete units, on storage undergo a sintering process which yields a coherent solid block which prevents convenient partitioning of the product.

The general process of sintering particularly as applied to metallurgical systems is well known and the mechanisms prevailing in a one-component system have been extensively studied.[2] Its application to ice has been described by Hobbs and Mason[20] and Jellinek and Ibrahim.[21] The latter authors investigated the rate of decrease in surface area of a powder consisting of ice spheres with radii of about 0.5 μm. This rate was studied as a function of temperature from −8 °C to −35 °C and the surface areas were measured by the BET method. When the rate constants for the changes were plotted (Fig. 8), they showed a remarkable change at −12.5 °C. Jellinek and Ibrahim believe that in this instance the sintering process is a plastic flow phenomenon caused by stresses due to interfacial tension forces. Viscosity values for ice as functions of temperature and stress were obtained in agreement with accepted values. In contrast, Hobbs and Mason observed the growth of a neck between two ice spheres under the microscope; their spheres were considerably larger (diameters from 50–700 μm). Hence quite different orders of magnitude for stresses are involved. Their results could be accounted for by an evaporation-condensation process, but Jellinek and Ibrahim showed that if the same mechanism were invoked for the much smaller ice spheres employed by them, a rate of reaction vastly greater than that observed experimentally would be expected. Somewhat surprisingly, the sharp discontinuity at −12.5 °C reported by Jellinek is not observed by Hobbs and Mason. There is clearly room here for further investigations, particularly of actual particulate frozen products and the accumulation of evidence as to whether −12.5 °C is critical to the storage of frozen products in practice.

In conclusion it is evident that although frozen foods and frozen food

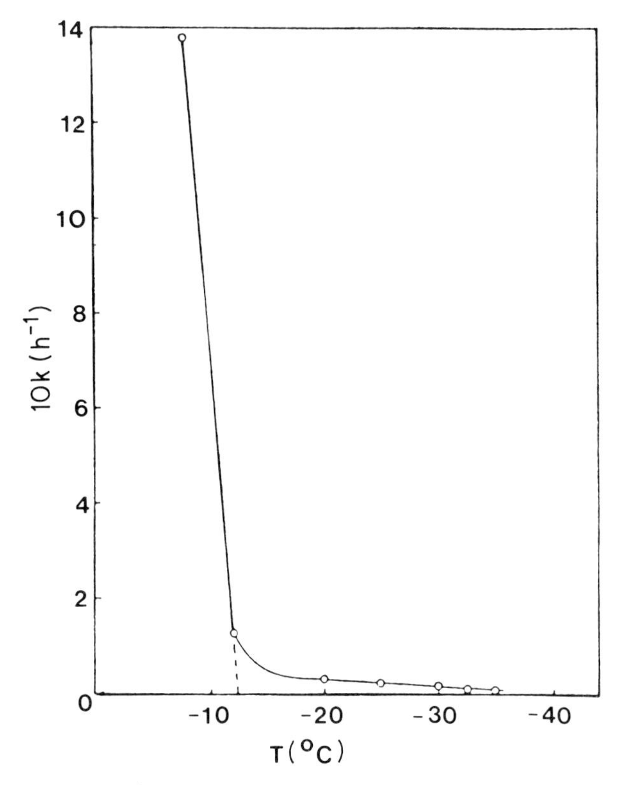

Fig. 8 Rate constants for decrease of surface area of powdered ice as a function of temperature. A rapid increase in the magnitude of the rate constants occurs at −12.5 °C

technology have been with us for over fifty years, there are still many basic problems of a physicochemical nature to be overcome and at the same time some most promising leads which could permit a major increase in the overall use of low temperatures in the preservation and storage of biological tissues and food products at low temperatures.

References

1. Angell, C. A. (1982). Water—A Comprehensive Treatise, Vol. 7, ed. F. Franks, pp. 1–82. Plenum Press, New York.
2. Ashby, M. F. (1974). *Acta Metallurgica* **22**, 275–285.
3. Berger, K. G., Bullimore, B. K., White, G. W. and Wright, W. B. (1972). The Structure of Ice Cream. Dairy Industries, **37**, 419–425.
4. Boutron, P. (1986). *Cryobiology* **23**, 88–102.

5. Brown, R. A., Yeh, Y., Burcham, T. S. and Feeney, R. E. (1985). *Biopolymers* **24**, 1265–1270.
6. Calvelo, A. (1981). Developments in Meat Science, Vol. 2, ed. R. A. Lawrie, pp. 125–158. Applied Science Publishers, London.
7. DeVries, A. L., Komatsu, S. K. and Feeney, R. E. (1970). *J. Biol. Chem.* **245**, 2901–2913.
8. DeVries, A. L. and Wohschlag, D. E. (1969). *Science* **163**, 1074–1075.
9. Diller, K. R. (1985). *Cryobiology* **22**, 268–281.
10. Dufour, L. and Defay, R. (1963). Thermodynamics of Clouds. Academic Press, New York.
11. Fennema, O. (1975). *Water Relations of Foods*, ed. R. B. Duckworth, pp. 539–556. Academic Press, London.
12. Fletcher, N. H. (1970). The Chemical Physics of Ice. Cambridge University Press, Cambridge.
13. Fahy, G. M., MacFarlane, D. R., Angell, C. A. and Meryman, H. T. (1984). *Cryobiology* **21**, 407–426.
14. Franks, F. (1982). *Water—A Comprehensive Treatise*, Vol. 7, ed. F. Franks, pp. 215–338. Plenum Press, New York.
15. Franks, F. (1985). Biophysics and Biochemistry at Low Temperatures. Cambridge University Press, Cambridge.
16. Franks, F., Mathias, S. F. and Trafford, K. (1984). *Colloids Surfaces* **11**, 275–285.
17. Franks, F., Darlington, J., Schenz, T., Mathias, S. F., Slade, L. and Levine, H. (1987). *Nature* **325**, 146–147.
18. Farrant, J. (1969). *Nature*, **222**, 1175–1176.
19. Hatley, H. M., Franks, F. and Day, H. (1986). *Biophysical Chemistry*, **24**, 187–192.
20. Hobbs, P. V. and Mason, B. J. (1964). *Phil. Mag.* (8th series) **9**, 181–197.
21. Jellinek, H. H. G. and Ibrahim, S. H. (1967). *J. Colloid Interface Sci.* **25**, 245–254.
22. Krog, J. O., Zachariassen, K. E., Larsen, B. and Smidsrod, O. (1979). *Nature*, **282**, 300–1.
23. Levine, H. and Slade, L. (1986). *Carbohydrate Polymers*, **6**, 213–244.
24. Lindow, S. E. (1983). *Ann. Rev. Phytopathol.* **21**, 363–384.
25. MacFarlane, D. R., Kadiyala, R. K. and Angell, C. A. (1983). *J. Chem. Phys.* **79**, 3921–3927.
26. Mathias, S. F., Franks, F. and Hatley, R. H. M. (1985). *Cryobiology* **22**, 537–547.
27. Muhr, A. H. (1983). The influence of polysaccharides on ice formation in sucrose solutions. PhD thesis, University of Nottingham.
28. Muhr, A. H., Blanshard, J. M. V. and Sheard, S. H. (1986). *J. Fd Technol.* **21**, 587–603.
29. Muhr, A. H. and Blanshard, J. M. V. (1986). *J. Fd Technol.* **21**, 683–710.
30. Rasmussen, D. H. and Mackenzie, A. P. (1972). Water Structure at the Water–Polymer Interface. Ed. H. H. G. Jellinek, Plenum Press, New York.
31. Reid, D. S., Foin, A. T. and Lem, C. T. (1985). *Cryoletters* **6**, 189–198.

5 Fat Crystallization

P. WALSTRA

Department of Food Science, Wageningen Agricultural University

1 INTRODUCTION

Crystallization of lipids is a vast and complicated subject. An authoritative and up-to-date discussion of the fundamentals is given by Larsson.[12] Here, we will restrict ourselves to one class of lipids, the triacyl glycerides, called fats for short. An exhaustive, but now somewhat outdated treatment is that by Bailey.[1]

The crystallization of fat may affect several properties of a food:

1. The *consistency* of butter, margarine, shortening, chocolate, heavy cream, etc. This will be briefly discussed below.
2. The *mouthfeel* of a high-fat food. Large fat crystals may give a sandy or grainy feeling. The melting of a considerable quantity of crystalline fat in the mouth imparts a cool impression (particularly in butter). If many small crystals remain in the mouth (i.e. do not melt), the product appears sticky.

FOOD STRUCTURE AND BEHAVIOUR
ISBN 0-12-104230-8

3. *Segregation.* Crystals formed in an oil may sediment and a plastic fat with too few or too large crystals will exhibit "oiling off".
4. *Emulsion stability.* In a water-in-oil emulsion (like margarine), fat crystals adhering to the water droplets tend to stabilize these against coalescence (the so-called Pickering stabilization); the droplets also may be immobilized in a network of flocculated fat crystals. Fat crystals in the droplets of an oil-in-water emulsion tend to promote (partial) coalescence; see Chapter 6 of this book.

Several of these aspects are discussed by Mulder and Walstra,[13] who also summarize fat crystallization (pp. 33–53). Fat crystallization and its consequences are often interlinked with the technology of the isolation, purification and modification of the fat and, thereafter, of its manufacture into products. The most complete treatment of these matters is by Swern.[20]

2 MELTING RANGE

Natural fats contain different triglycerides that may range in melting point from −40 to +72 °C. Consequently, natural fats have a melting range which frequently spans several decades of temperature °C. The constituent fatty acids of a natural fat usually number at least ten but may be several hundred. If n different fatty acids are present, the possible number of different triglycerides is n^3. The number of triglycerides present in significant quantities, ranges from, say, 5 to 500. In some fats (e.g. milk fat) even the most abundant triglyceride makes up only one or two mole percent of the mixture. Extensive information about the composition of natural fats is given by Hilditch and Williams[7] and, more up-to-date but less extensive data by Sheppard *et al.*[17]

The melting point of any triglyceride, and thereby the melting range of a mixture of triglycerides, depends on its fatty acid residues. Table 1 gives examples of melting points of fatty acids. The factors which affect the melting points are:

Chain length. The longer the chain the higher the melting point. However, fatty acids with an odd number of carbon atoms on average melt 5 K lower than even numbered ones.

Number of double bonds. The more unsaturated, the lower the melting point. Consequently, hydrogenation of a fat, i.e. saturating (a fraction of) the double bonds, shifts its melting range to higher values; consequently, hydrogenation is often called "hardening". Incidentally, hydrogenation also causes shifts in the position of double bonds and *cis-trans* isomerization.

Table 1 Melting points (°C) of some fatty acids and triglycerides if crystallized in the stablest modification

			°C
Fatty acids	4:0	butyric	−8
	10:0	capric	31
	16:0	palmitic	63
	18:0	stearic (S)	70
	18:1, *cis*	oleic (O)	16
	18:2, *cis*	linoleic	−5
	18:3, *cis*	linolenic	−14
	18:1, *trans*	elaidic (E)	44
Triglycerides		SSS	73
		SOS	42
		SSO	38
		SOO	24
		OOO	5
		SES	60

Double bond stereoisomerism. Trans double bonds give a far smaller lowering (20–30 K) of the melting point than *cis* ones. This is because a *trans* double bond permits the acyl chain to assume an almost linear conformation, while a *cis* configuration necessarily causes a kink.

Position of double bonds. This may make a difference of up to 20 K. Conjugated *cis* double bond pairs permit an almost linear chain conformation and thus give a 25 K higher melting point than non-conjugated pairs.

Branching of the chain decreases the melting point by 1–40 K.

Moreover, the positioning of the fatty acids over the triglycerides greatly affects the melting range and those with a more uneven and asymmetric distribution of the different fatty acid residues tend to be lower melting. Examples are given in Table 1. Few natural fats have a random distribution of residues over the glyceride positions and inter-esterification, which leads to a random distribution, consequently causes a change in melting range, usually extending it to higher temperatures. For example, inter-esterification of cottonseed oil increases its final melting point or clear point (i.e. the temperature at which the last crystals disappear when slowly heating the fat) from 10 to 34 °C. On the other hand, the clear point of lard is little affected by inter-esterification (though other characteristics of crystallization are), while a mixture of 25% (impure) tristearate and 75% soybean oil showed a clear point of 60 °C before and of 27 °C after inter-esterification. The latter change is easily explained by the following points.

The melting range of a fat is not just the composite of the melting points of the component triglycerides. This is because the higher melting components dissolve in the liquid fat. For a mixture of two triglycerides that are fairly similar but differ appreciably in melting point, the solubility x (expressed as mole fraction) of the higher melting component at temperature T is given by[5]

$$R \ln x = \Delta \bar{H}\left(\frac{1}{T_f} - \frac{1}{T}\right) \tag{1}$$

where $\Delta \bar{H}$ is its molar heat of fusion (for many triglycerides between 100 and 200 kJ mol^{-1}) and T_f its melting point (in K); R is the gas constant (8.3 J mol^{-1} K^{-1}). For example, a fat in which tristearate is the highest melting triglyceride ($\Delta \bar{H} = 205$ kJ mol^{-1}, $T_f = 73$ °C), and of which the clear point is 37 °C, would yield $x = 3 \times 10^{-4}$. Figure 1 contains the calculated melting curve of an equimolar mixture of tristearate and trioleate; such calculations agree very well with experimental results. However, in more complicated mixtures, deviations from equation 1 are observed.[10] If we revert to the mixture of tristearate and soybean oil, inter-esterification would reduce x from ~0.25 to ~0.015; hence the far lower clear point.

Figure 1 also gives examples of a part of the melting curves of some natural fats. It should be realized that for each type of fat different samples may yield widely different melting curves for the following reasons:

The composition of the triglyceride mixture may appreciably differ between samples of different origin.

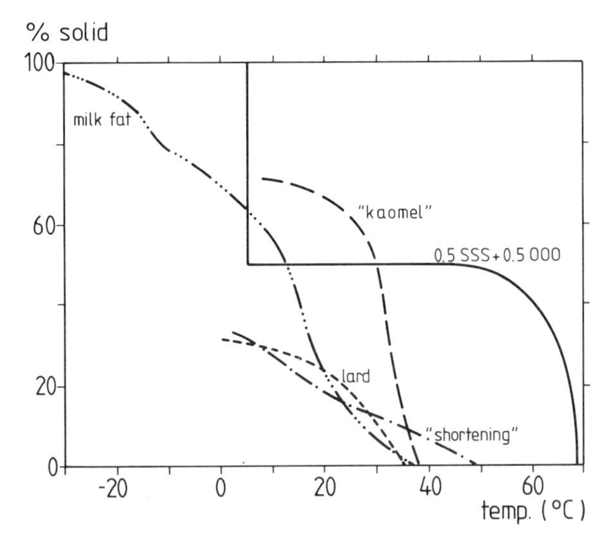

Fig. 1 Examples of melting curves (determined by the author). Milk fat, lard, a cocoa butter imitation (kaomel) and a partially hydrogenated vegetable oil (shortening). Also given is a calculated curve for an equimolar mixture of tristearin and triolein (β modification)

Technological treatment may affect fat composition.
As will be seen below, other factors than triglyceride composition, e.g. temperature history, may affect the melting curve.

3 POLYMORPHISM

Triglycerides, like other molecules with long aliphatic chains can crystallize in different polymorphic modifications (see Refs. 3 and 12 for reviews). The main difference is the chain packing, which can be hexagonal, orthorhombic or triclinic, these modifications being designated α, β' and β, respectively. These chain packings are characterized by their X-ray diffraction patterns (the so-called short spacings that represent the distances between chains) and by their ir-spectra. Figure 2 gives examples of the chain packing. Table 2 gives some properties of the different polymorphs for tristearate; it is evident that density, enthalpy of fusion and melting point increase in the order α, β', β. In the α form, the density is comparatively low and the chains have considerable rotational freedom. Presumably, a liquid triglyceride at temperatures not much above the α melting point exhibits appreciable short-range order, with a chain packing somewhere between tetragonal and hexagonal[11] as suggested in Fig. 3. Consequently, the transition from the liquid to the α form is not one from a completely disordered to a perfectly ordered state.

Figure 4 gives examples of the packing of trilaurin molecules in α and β crystals. Actually, several different packing modes are possible within each of the β' and the β modification[8,12] and Fig. 5 gives an example due to double and triple chain packing; other differences are in the tilt of the chains. Consequently, numerous different crystal lattices can occur, according to chain lengths and unsaturation of the component fatty acids. Nevertheless, for many purposes, the classification into α, β' and β suffices.

These are monotropic modifications, i.e. there is only one stable form (usually β) and the others are unstable. The possible transitions are given in Fig. 6, other transitions cannot occur. This implies (see e.g. Table 2) that a triglyceride may show a double or even a triple melting point, and this can

Table 2 Some properties of the three main polymorphic modifications of tristearin

Property	Modification α	β'	β
Main short spacings (nm)	0.415	0.38 and 0.42	0.46
Melting point (°C)	55	64	72
Enthalpy of fusion (J g^{-1})	163	180	230
Melting dilatation (cm^3 kg^{-1})	119	131	167

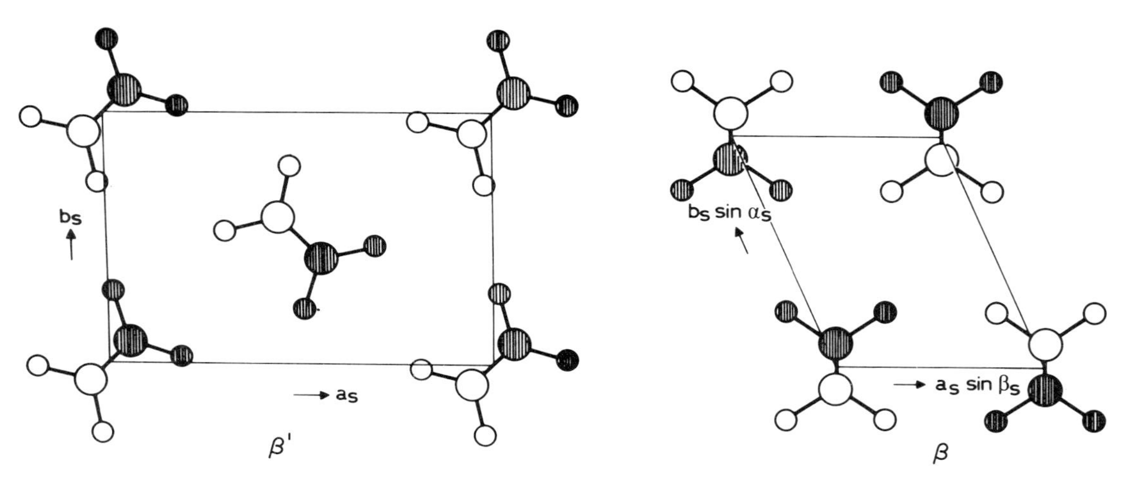

Fig. 2 Two different chain packings of hydrocarbon chains; orthorhombic, corresponding to the β' modification; and triclinic, corresponding to the β modification (from ref. 8)

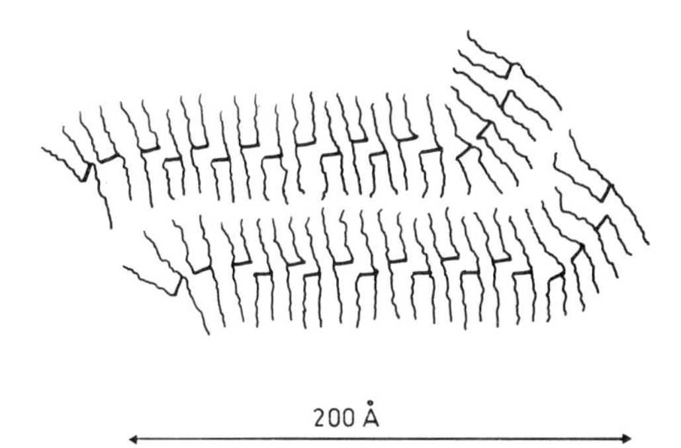

Fig. 3 Short range order in a liquid triglyceride at a temperature close to the melting point (from ref. 11)

indeed be observed if the liquid triglyceride is cooled rapidly and subsequently heated. There are a few exceptions to the scheme in Fig. 6 in that for some triglycerides the β' form is the stablest. The α modification is formed most readily, but it is usually very unstable; hence it rapidly (within minutes) transforms into the β' form. The latter may remain a bit longer (e.g. an hour), but eventually is transformed into β. These observations on lifetime pertain to fairly pure triglycerides; in a multicomponent fat the unstable polymorphs may live very much longer for reasons given in the next section. The presence of diglycerides may particularly enhance the lifetime of the β' form.[6] There is no agreement among authors whether the polymorphic transitions occur via the liquid state or not.

Most multicomponent fats also show polymorphism, as is illustrated in Fig. 7; after cooling briefly, the fat shows a double melting curve, which

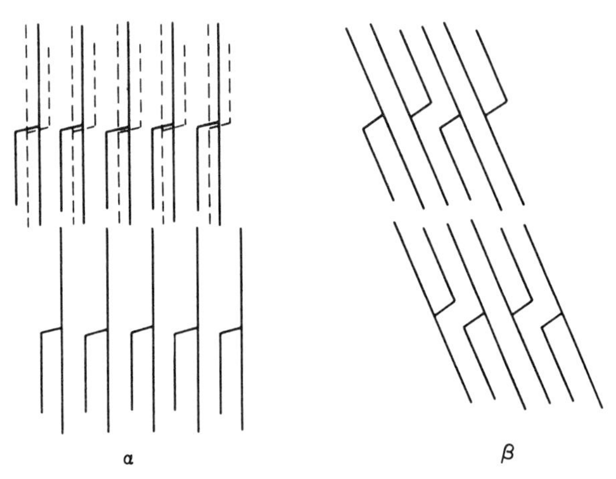

Fig. 4 Approximate packing of trilaurin molecules in the α and β modifications

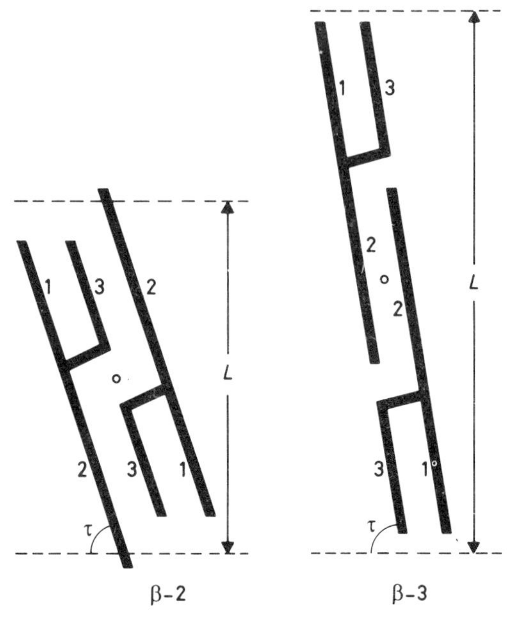

Fig. 5 Arrangement of triglyceride molecules in two different β modifications, β-2 and β-3 (from ref. 8)

feature disappears after keeping it somewhat longer at low temperature. Such storage thus leads to isothermal polymorphic transitions, but this does not imply that *all* crystals in a less stable polymorphic state have disappeared. For example, it was found in milk fat[2] and cocoa butter[5] that all these modifications were present and that they disappeared on heating to the following temperatures.

Modification	α	β'	β	
Milk fat	22	30	35	°C
Cocoa butter	24	28	35	°C

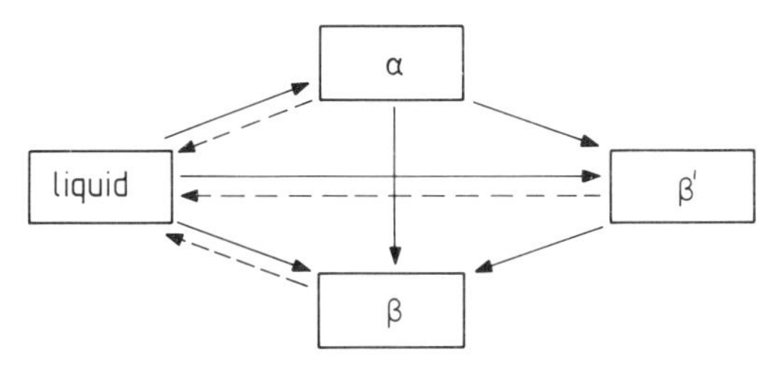

Fig. 6 Possible transitions between the liquid state and the various polymorphs of triglycerides; common case. —— exothermal, ---- endothermal transition

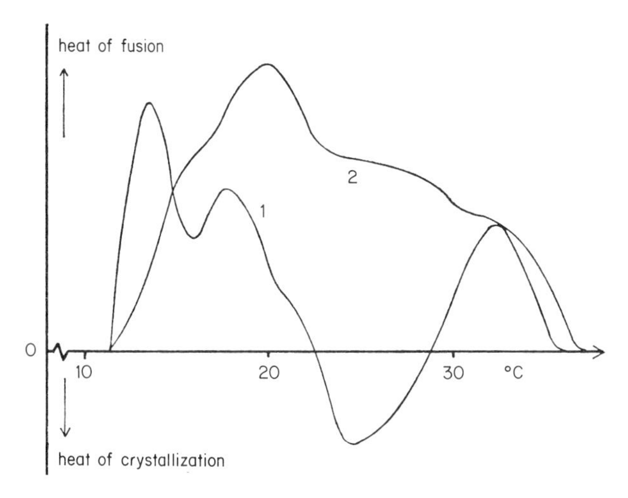

Fig. 7 Melting curves of milk fat, given as heat effects per °C rise, determined by differential thermal analysis. 1: fat rapidly cooled to and kept for 5 min at 11 °C, 2: the same but kept for 90 min at 11 °C (after ref. 2)

4 COMPOUND CRYSTALS

A compound or mixed crystal contains two or more different components (molecular species). In some the components may occur in all proportions (these are also designated solid solutions), in others the compositional range is restricted. Compound crystals are more likely to be formed if the different molecular species are more similar in shape, size and properties. Mixtures of triglycerides easily form compound crystals in the α modification; since the molecular packing is not very dense, there is some freedom of fitting different molecules in one crystal lattice. Compound crystals are thus less easily formed in the β' and are very limited in the β modification. Most compound fat crystals are not stable but tend to slowly rearrange into purer crystals, which often is accompanied by a polymorphic transition. Compound crystals have a lower enthalpy of fusion and a lower density than pure crystals of the same modification.

Fats with a wide compositional range are particularly prone to crystallize in the compound form, and even more so when the temperature is lower. A general explanation for such behaviour is a consequence of the fact that for any of the numerous components present the supersaturation may be low, but for a group of similar components the aggregate supersaturation may be considerable. By the same token the unstable polymorphs may be rather persistent. For pure triglycerides, the β form is rapidly attained, but in some binary mixtures that form compound crystals the β' form may doggedly persist.[4,15] In most partially hydrogenated vegetable oils, some crystalline material in the β' form persists almost indefinitely (at least at a fairly low temperature), but the α form rapidly disappears.[5] In a multicomponent fat, such as milk fat, even the α form may persist, almost indefinitely.[2]

Mixtures of two or three triglycerides show a wide range of phase diagrams, according to composition and modification.[10,16] The simplest type, two components giving compound crystals over the full compositional range, is illustrated in Fig. 8. Above the liquidus line there is only liquid, below the solidus there is only solid fat, and in between there are both. For instance, cooling a mixture with composition a_3 rapidly from T_1 to T_2 gives crystals with composition a_5, and a liquid phase a_2 remains. The mass ratio of solid to liquid is $(a_3 - a_2)/(a_5 - a_3)$. When the temperature is changed, the composition of both liquid and solid phase will change, but it may take a long time before equilibrium is reached; this is because rearrangement has to take place in the crystalline phase, which is a slow process, the more so as the temperature is lower. Further cooling, for instance to T_3, will leave the solid phase a_5 largely as it is, while the liquid phase a_2 will segregate into a_1 (liquid) and a_4, but the amount of a_1 decreases in this process; hence the proportion of solid fat increases.

In a natural fat the situation is more intricate than that of the binary mixture of Fig. 8. There are not two but many components; some give compound crystals and others do not; polymorphism is involved. There is certainly more than one type of compound crystal. But there may be series of many compound crystals, existing side by side over most of its melting range.

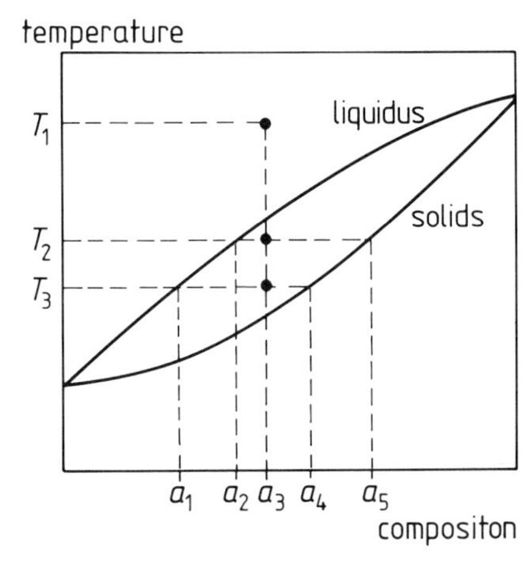

Fig. 8 Hypothetical phase diagram of a binary mixture, giving compound crystals over the full compositional range

In such a case qualitative conclusions can be drawn from a phase diagram like Fig. 8, and such conclusions are mostly born out by experimental results.[13]

If extensive compound crystallization occurs it has several consequences:

1 The melting range is narrowed.
2 The temperature at which most of the fat melts depends on the temperature at which the crystallization took place. Cooling liquid fat in two steps within the crystallization range may give rise to two melting maxima.
3 Cooling in steps, or cooling very slowly, gives less crystalline fat than rapid cooling to the lowest temperature.
4 Precooling to a lower temperature before bringing to the final temperature results in more crystalline fat than directly cooling to the latter; see e.g. Fig. 9*A*.
5 Unstable polymorphs persist longer and other slow arrangements occur; this is discussed later.

5 NUCLEATION

Nucleation commonly occurs in the α modification, often as compound crystals. At temperatures above the final melting point of the α modification nucleation may occur in another polymorphic form, albeit at a much slower rate. Homogeneous nucleation occurs at a measurable rate at a temperature only 20–25 K below the final α melting point,[19] hence roughly 35 K below the final melting point of the fat. But nucleation is usually heterogeneous. As soon as some fat crystals have formed, they act as catalytic impurities for fat crystals of other composition, a clear case of epitaxy. This implies that little hysteresis between crystallization and melting curves occurs, as is illustrated in Fig. 9*A* (the hysteresis at low temperature is due to compound crystallization, as discussed above).

Most natural fats contain many catalytic impurities for heterogeneous nucleation. Most of the impurities presumably consist of micelles of monoglycerides; increased lipolysis usually leads to enhanced nucleation.[18,23] The number of impurities suffices to initiate crystallization in a bulk fat, even at a few degrees below the final meeting point. If the fat is finely divided, as in an emulsion, the number of catalytic impurities per unit volume (N) may be far too low to produce nuclei in every emulsion droplet, and considerable supercooling may occur: Fig. 9*B*. If the droplet volume is v, the maximum fraction of the volume of an emulsified oil containing impurities and thus eventually crystals is

$$y_m = 1 - \exp(-vN) \tag{2}$$

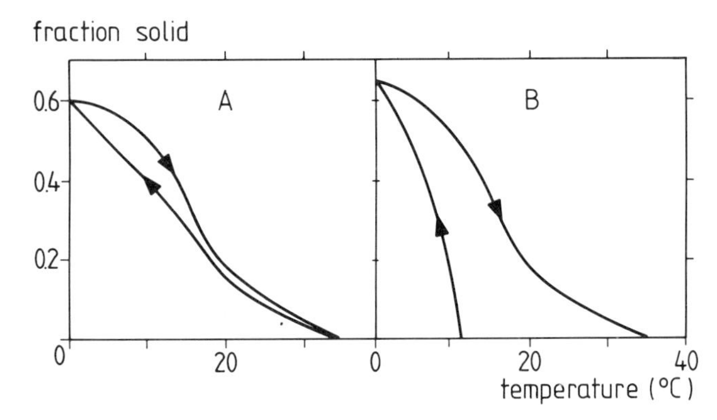

Fig. 9 Hysteresis between cooling and heating curves for milk fat. Cooling at each temperature for 24 h. Heating curve obtained by heating the sample kept for 24 h at 0 °C to the higher temperatures for 30 min. *A*: Fat in bulk. *B*: Fat finely emulsified (average droplet volume ~0.8 μm³) (after ref. 23)

N always increases with decreasing temperature T and a common empirical rule is that log N is linear with T, N for example doubling for every 1.5–2 K lowering of temperature. To give a rough idea, N may be 1 μm^{-3} at 30 K below the final melting point, which is also roughly the homogeneous nucleation temperature. Therefore, where a fat is finely divided, not only the probability of nucleation (hence crystallization) occurring at all in a droplet may be small, but also the nucleation rate is affected and it may take hours before nucleation is completed at a temperature which is only 20 K below the final melting point. The kinetics of nucleation in emulsified fat have been worked out by Walstra and Van Beresteijn.[23]

Thus far we have considered primary nucleation, but in most fats copious secondary nucleation occurs; that means that many crystal nuclei spring up in the vicinity of a crystal once it has been formed. In consequence the majority of fat crystals remain small, even when the rate of primary nucleation is low.

6 RATE OF CRYSTALLIZATION

Fat crystallization is usually a very slow process and the reasons for this are not difficult to see:

—The triglyceride molecule has three long, flexible chains and fitting these into a crystal lattice takes considerable time. This is clearly demonstrated when comparing the rates of crystal growth and dissolution under comparable conditions (see Fig. 10).

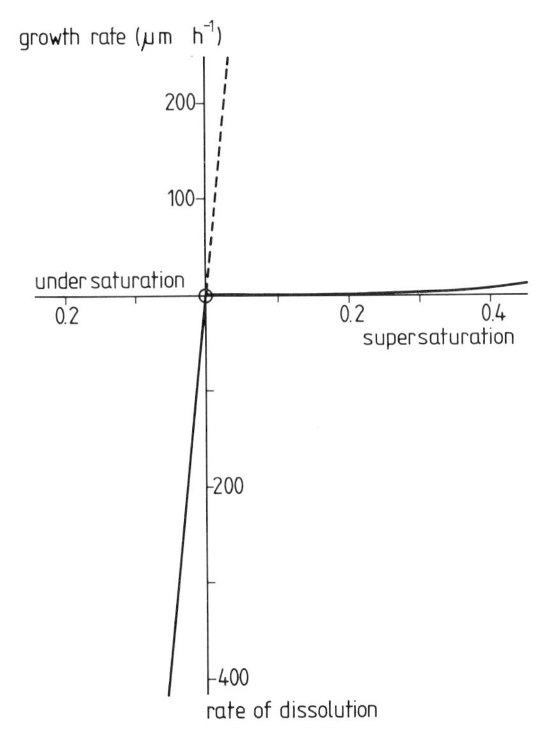

Fig. 10 Rate of growth and dissolution of a certain crystal face of a trilaurin crystal in oil at various supersaturation or undersaturation, expressed as $(C - C_{sat})/C_{sat}$. —— experimental, ---- extrapolated (after ref. 19)

—If multi-component fat is involved, the supersaturation of each component is low.

—Many different triglyceride molecules almost fit into the same crystal lattice and the presence (often abundant) of nearly fitting molecules may considerably retard the incorporation into the lattice of the perfectly fitting ones ("competitive inhibition"). If no compound crystals were formed, this with the previous consideration would even more strongly reduce the already slow rate of crystallization.

—Competitive inhibition may also be caused by other substances, e.g. diglycerides and monoglycerides.[22]

—As soon as crystals have formed, the consistency of a fat becomes such as to greatly hinder the efficient removal of heat, including the considerable heat of crystallization (mostly between 100 and 200 kJ per kg crystals). This does not necessarily apply to an emulsified fat.

The isothermal crystallization rate, of course, greatly depends on temperature. In a multi-component fat the initial rate may, for instance, double for each 5 K lowering of temperature, but it may take a very long time for the crystallization to be more or less complete, perhaps 24 h (see Fig. 11 and the discussion in the next section). The following factors may be responsible for the effect of temperature on the rate of crystallization expressed as the mass of crystals formed per unit time (dc/dt):

—At a lower temperature a greater number of crystals is usually formed which yield a greater surface area.

—At a lower temperature more triglycerides can form crystals. This would not affect the relative crystallization rate $d \ln c/dt$.

—For each triglyceride, however, the supersaturation steeply increases with decreasing temperature, at least within a certain temperature range (see equation 1 and Fig. 1). Since dc/dt is roughly proportional to the cube of the supersaturation,[19] the effect on overall crystallization rate is very considerable.

—At a lower temperature, more triglycerides can crystallize in the β' and still more in the α modification, which implies an easier fitting into the crystal lattice and more compound crystallization (hence a higher effective supersaturation and less competitive inhibition).

—At lower temperature the existence of an activation free energy for crystallization (which is presumably for the most part due to the fitting difficulty of the triglyceride molecules) would cause a slower rate of crystallization. Obviously, the effects previously mentioned are overriding.

Admittedly, most of what we can say about the rate of crystallization of fats is only of a qualitative nature.

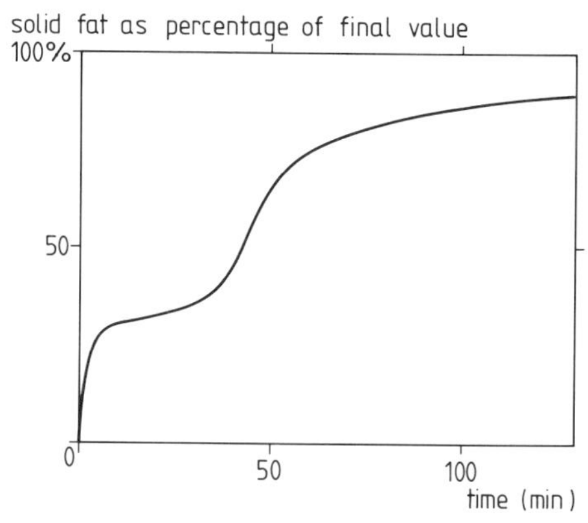

Fig. 11 Example of isothermal crystallization of milk fat at 11 °C as a function of time after bringing to 11 °C

7 RECRYSTALLIZATION

In a sense, crystallization may never become complete in a multi-component fat, because thermodynamic equilibrium is never reached and the reshuffling of molecules between various crystals may continue on. In fact, such a reshuffling already occurs during the initial crystallization, which is the reason why curves of the percentage crystallized versus time often exhibit inflection points (see e.g. Fig. 11). The main changes are:

—Compound crystals change in composition, generally segregating into purer ones. Changes will always occur if the temperature is varied.
—Unstable polymorphs (α and β') are transformed into more stable forms (β' and β). This and the previous change may often occur simultaneously. Hence, most polymorphic transitions probably proceed via the melt.
—"Ostwald ripening" occurs, i.e. large crystals grow at the expense of smaller ones. Since a relative supersaturation of at least 0.1 is needed to give measurable crystallization (see Fig. 10), the critical radius of curvature calculated from the Kelvin equation, is well below 0.1 μm.

The rate of recrystallization greatly depends on temperature. If it is so low that a considerable fraction (say >0.7) of the fat is crystalline, changes are slow; the crystal composition is more or less "frozen". Generally speaking, at a higher temperature it takes a shorter time to obtain a near-equilibrium situation, although the initial crystallization rate is much slower. A temporary temperature rise may accelerate recrystallization. The quantity, composition and size of crystals in a multi-component fat are therefore all dependent on its temperature history.

8 SIZE AND ARRANGEMENT OF CRYSTALS

Most fat crystals are small, particularly when the fat has been cooled rapidly; presumably, the extensive secondary nucleation is responsible for the large number and therefore small size of the crystals. Many crystals are too small to be visible with a light microscope, a problem which is exacerbated by their shape. Many fats tend to form thin and fairly long, platelet-shaped crystals. Crystals of more than a few μm long are rarely observed in a rapidly cooled fat in the first few days after cooling. There are, however, considerable differences between fats in size and shape of crystals; the factors governing those properties are poorly understood, though the effect of some surfactants on the growth rate of the faces of particular crystals has been determined.[22]

As soon as the crystals formed exceed a certain size (very roughly 0.1 μm) they flocculate, because of the net Van der Waals' attraction between them, and virtually no repulsion (either electrostatic or steric), except at atomic distances.[21] Unless the crystal mass is very small, the flocculating crystals form a network extending throughout the whole volume, thereby giving the mixture the properties of a solid. A 10% crystalline fat is usually more than sufficient to achieve this. The fat thus acquires an elastic modulus; this only applies for very small deformations ($\ll 1\%$). If larger deformations are imposed, the fat begins to yield under the stress, as bonds in the network are broken, and to flow. The critical stress is called the yield value or yield stress.

If the fat is kept for some time, the elastic modulus and yield stress increase considerably. The explanation must be that ongoing crystallization and particularly recrystallization cause a growing together or sintering of the flocculated crystals, thereby very much enhancing the strength of the bonds in the network. The sequence of events leading to this situation is depicted in Fig. 12. The elastic modulus of the fat may now be of the order of 10^6 Pa and the yield stress 10^7 Pa.[6]

Working or kneading, i.e. strongly deforming the fat, breaks many bonds in the network, and the fat becomes much softer (so called "work softening").

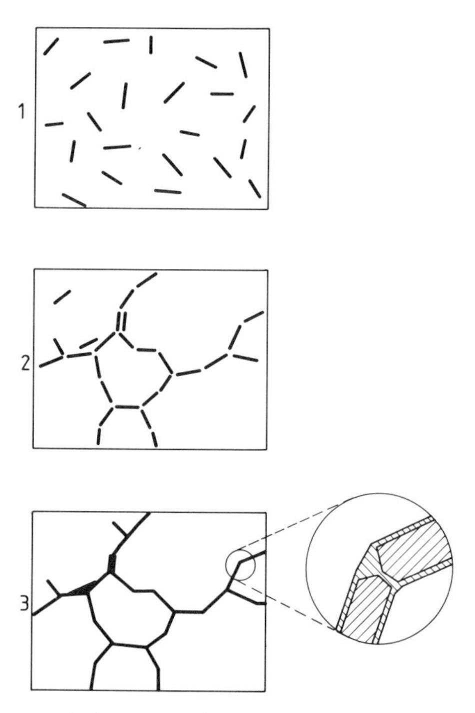

Fig. 12 Various stages during crystallization showing flocculation and sintering of fat crystals. Highly schematic

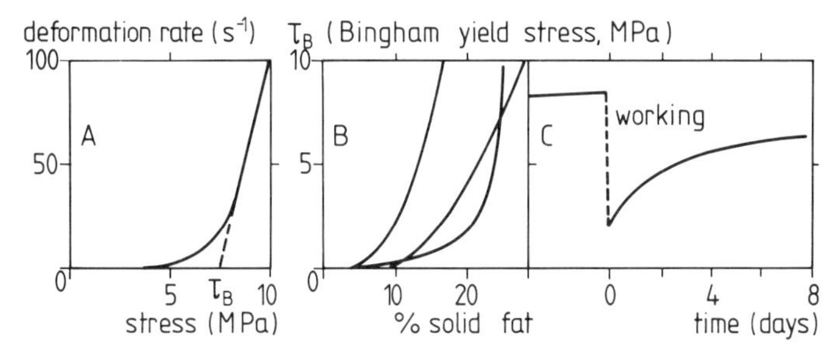

Fig. 13 Rheological properties of a partially crystallized fat. *A*: Deformation rate as a function of stress applied and definition of the Bingham yield stress τ_B. *B*: Relations between τ_B and % solid for different fats. *C*: τ_B as a function of time and of intensively working the sample (at $t = 0$). Highly schematic examples

When such processes cease the crystals start to flocculate and even sinter again, because of recrystallization. In this situation, Ostwald ripening may significantly promote the recrystallization, since the fracture of sintered crystals will leave sharp edges; the resultant very small radius of curvature thermodynamically favours local dissolution of crystalline material. The ensuing effects on the firmness of the fat are depicted in Fig. 13. Temperature fluctuation, particularly a temporary rise in temperature, predominantly enhances firmness (by up to a factor of 4), since it promotes sintering.

In some fats, recrystallization causes the formation of very large crystals, particularly if the recrystallization is slow (e.g. over several weeks). It usually involves the $\beta' \rightarrow \beta$ transition, leading to the formation of large (up to 0.1 mm diameter) spherulite crystals. Such a fat may become very soft.

All in all, the firmness of a fat may, for the same proportion of crystalline material, vary by almost an order of magnitude.

9 CRYSTALS IN DROPLETS

A fat may be finely divided, as in an oil-in-water emusion. In most emulsions, the average droplet diameter is between 0.3 and 5 μm, which implies that 1 g of fat is spread over 10^{10}–10^{14} separate droplets. Much coarser dispersions occur as well; for instance, fat cells in animal tissues may be as large as 100 μm. The dispersion of the fat may have several consequences:

—The supercooling needed for nucleation to occur may be more severe. This has been discussed already. If the surface of the droplets is covered by a surfactant with long aliphatic chains, heterogeneous surface nucleation may occur at the droplet boundary; this may cause nucleation at an appreciable rate at a temperature some degrees above the homogeneous nucleation temperature.[18]

—The crystal composition, i.e. predominant polymorphs and composition of compound crystals, may differ from that in bulk fat. An important cause may be that crystallization in a small droplet can be effectively isothermal, while this can not be achieved in large mass of a multi-component fat.

—Crystal size and arrangement may differ. Generally, crystals cannot grow larger than the droplets, though in exceptional cases crystals growing out of the droplets have been observed. Usually the crystals are much smaller than the droplet and subsequently they flocculate into a network but with a yield stress of some 10^4–10^5 Pa, which is lower than that of a bulk fat.

—Under some conditions, some crystals become oriented at the droplet boundary, where they are tenaciously held by surface forces (see Chapter 6).

—Since crystals in one droplet cannot form a continuous network with those in another droplet, the rheological properties of an oil-in-water emulsion of 50% fat in which part of the fat is crystallized is very different from that of a water-in-oil emulsion of the same composition; the former system is a liquid (with a viscosity about 15 times that of water); the latter has a plastic consistency with a considerable yield stress.

—The stability of oil droplets to (partial) coalescence and disruption is usually greatly influenced by the presence of crystals in the droplets (see Chapter 6).

10 ESTIMATION OF SOLID FAT CONTENT

The classical methods of estimating the proportion of a fat that is crystalline (solid), employ either the heat of fusion (calorimetry, DTA, DSC) or the melting dilatation of the crystals (dilatometry). However, these parameters are by no means invariant and therefore wholly reliable:

—they differ, with the melting point of the triglyceride and therefore throughout the melting range (e.g. the heat of fusion of tristearin is 220; tricaprin, 170 and triolein, 110 J g^{-1});

—they thus also differ with the polymorphic modification (see Table 2);

—the existence of compound crystals influences (lowers) the values.

A better method is to determine the proportion of liquid by means of wide-line proton NMR, employing either continuous wave or a pulse mode. Even so, it is important to realize that the solid fat content usually depends on temperature history and that the latter in practice is hard to reproduce in the laboratory.

References

1. Bailey, A. E. (1950). Melting and Solidification of Fats. Interscience, New York.
2. Beresteijn, E. C. H. van (1972). *Neth. Milk Dairy J.* **26**, 117–130.
3. Chapman, D. (1962). *Chem. Rev.* **62**, 433–456.
4. Haighton, A. J. (1963). *Fette Seifen Anstrichmittel* **65**, 479–482.
5. Hannewijk, J. (1964). *Chemisch Weekblad* **60**, 309–316.
6. Hernqvist, L. and Anjou, K. (1983). *Fette Seifen Ansrichmittel* **85**, 64–66.
7. Hilditch, T. P. and Williams, P. N. (1964). The Chemical Constitution of Natural Fats, 4th edn., Chapman & Hall, London.
8. Jong, S. de (1980). Triacylglycerol crystal structures and fatty acid conformation—a theoretical approach. Thesis, State University of Utrecht.
9. Knoester, M., de Bruyne, P. and van den Tempel, M. (1968). *J. Crystal Growth* **4**, 776–780.
10. Knoester, M., de Bruijne, P. and van den Tempel, M. (1972). *Chem. Phys. Lipids* **9**, 309–319.
11. Larsson, K. (1972). *Fette Seifen Anstrichmittel* **74**, 136–142.
12. Larsson, K. (1985). Lipid Handbook, eds F. B. Padley and F. D. Gunstone, Chapman & Hall, London.
13. Mulder, H. and Walstra, P. (1974). The Milk Fat Globule. Emulsion Science as applied to Milk Products and Comparable Foods. C.A.B., Farnham Royal and Pudoc, Wageningen.
14. Precht, D. and Greif, R. (1977). *Kieler Milchw. Forsch.ber.* **29**, 265–285.
15. Precht, D. and Greif, R. (1978). *Kieler Milchw. Forsch.ber.* **30**, 157–178.
16. Rossel, J. B. (1967). *Adv. Lipid Res.* **5**, 353–408.
17. Sheppard, A. J., Iverson, J. L. and Weihrauch, J. L. (1978). Handbook of Lipid Research. Vol. 1. Fatty acids and Glycerides, ed. A. Kuksis, 341–379, Plenum, New York.
18. Skoda, W. and van den Tempel, M. (1963). *J. Colloid Sci.* **18**, 568–584.
19. Skoda, W. and van den Tempel, M. (1967). *J. Crystal Growth* **1**, 207–217.
20. Swern, D. (Ed.) (1979, 1982). Bailey's Industrial Oil and Fat Products, 4th edn., Vols. I and II, Wiley, New York.
21. Tempel, M. van den (1961). *J. Colloid Sci.* **16**, 284–296.
22. Tempel, M. van den (1968). SCI-Monograph **32** (Surface-active Lipids in Foods), 22–33.
23. Walstra, P. and van Beresteijn, E. C. H. (1975). *Neth. Milk Dairy J.* **29**, 35–65.

6 Physical Principles of Emulsion Science*

P. WALSTRA

Department of Food Science, Wageningen Agricultural University

An emulsion is a dispersion of one liquid in another, both of which are mutually insoluble or only sparingly so. The preparation, properties and physical stability of emulsions involve several branches of science, notably colloid and surface science and hydrodynamics. In this paper we will confine ourselves to the problems that are of particular importance in edible emulsions.

1 ASPECTS

A food emulsion should fulfil the same quality criteria as any other food

* The greater part of this chapter has appeared earlier under the title "Physical Chemistry of Food Emulsions" in Volume 5 of the Proceedings of the 6th International Congress of Food Science and Technology (held in Dublin in 1983) and is reproduced here by kind permission of Boole Press, Limited, Dublin.

FOOD STRUCTURE AND BEHAVIOUR
ISBN 0-12-104230-8

product, being safe, palatable, durable, etc. Of such criteria, the following aspects are within the present subject.

A Appearance

This is principally a consequence of homogeneity, i.e. absence of segregation (hence physical stability) and colour. The whiteness of an emulsion increases with the scattering power of the droplets and therefore with the volume fraction (φ) of the disperse phase; it also is a function of droplet diameter (d), and is approximately maximal when $d = 1$ μm.[5]

B Flavour

Flavour components are usually present in both phases. When the emulsion is comminuted and diluted in the mouth, then if the taste substances are to be perceived, they must be released into the aqueous phase and similarly odour substances into the gaseous phase. These phenomena are very complicated; partition equilibria, diffusive and convective transport (in the bulk and through phase boundaries) and droplet coalescence, as well as the intricacies of perception all play a part. Some study has been made of release from oil droplets to the aqueous phase on dilution.[12] For water-in-oil emulsions it is qualitatively known that substances in the aqueous phase are better perceived if droplets are larger.[15]

C Rheological properties

Except for emulsions with a high volume fraction of disperse phase, rheological properties are comparable to those of other dispersions. The subject has recently been reviewed by Sherman.[20]

D Functional properties

These describe the behaviour of the emulsion during application. The ease with which an emulsion is diluted or mixed with composite foods greatly depends on whether it is oil-in-water (o/w) or water-in-oil (w/o). The latter type cannot be diluted with water. In such situations, rheological properties are also clearly important. For most applications stability to coalescence is needed, but there are other cases where a certain instability is desired. A special case is the whippability of an emulsion, a complicated subject outside the scope of this article; see for example Mulder & Walstra.[15]

E Stability

Some chemical and enzymic changes show particular features in emulsions, because of reactions occurring at the o/w interface. For example, it is known that enzymic hydrolysis and (chemical) autoxidation of fats is highly dependent on the properties of the absorbed layer at the interface[15] while microbial deterioration of w/o emulsions strongly depends on the droplet size

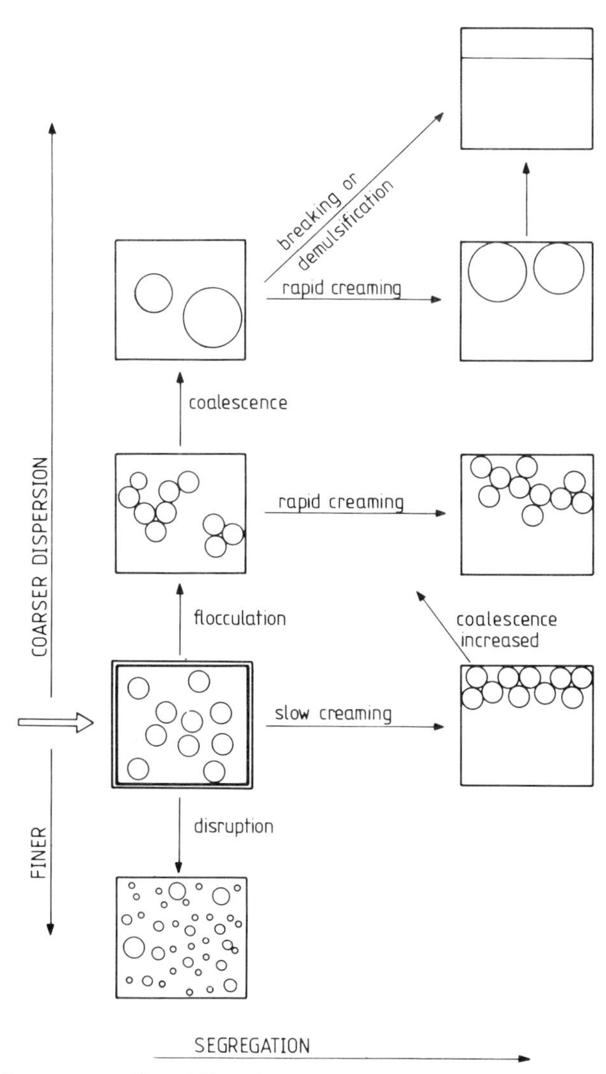

Fig. 1 Different types of instability of an o/w emulsion; simplified and highly schematic (from ref. 15)

distribution.[15,23] However, in this paper, we are primarily interested in physical stability. Such changes can usually be ascribed to creaming, flocculation, coalescence, or a combination of these phenomena (Fig. 1). The figure is, naturally, highly schematic, but it serves to show the interrelations.

It will be convenient if we not only confine our attention to this physical stability of emulsions but if we additionally impose the following limitations:

a *We only consider o/w emulsions.* In contrast, foods that are w/o emulsions (butter, margarine) have a complicated structure and their physical stability is primarily governed by other factors such as the size and location of fat crystals. A fairly recent development is the production of w/o/w emulsions[3,14] and though this may eventually open interesting possibilities in foods (separation of water-soluble components from each other) we are not aware yet of any practical applications.

b *The oil is a mixture of triglycerides.* Since triglycerides are effectively insoluble in water, this implies (for o/w emulsions) that there is no change in droplet size by isothermal distillation (so-called Ostwald ripening or disproportionation). It also implies that part of the droplets may crystallize, which has important implications for stability. An additional but coincidental factor is that the disperse phase nearly always contains some surface-active substances, such as monoglycerides, fatty acids or, possibly, phospholipids.

c *We do not consider microemulsions.* Their properties are very different from ordinary emulsions and there is little reason to think that their properties are significant in foodstuffs.

e *The continuous phase is liquid.* We therefore exclude products like sausage emulsions, in which a more or less solid matrix immobilizes small lumps of fat. This restriction implies that the emulsion droplets are spherical and of the order of a micrometer in diameter; they are covered by an adsorbed layer of surfactants, either small amphiphilic molecules, macromolecules (like protein) or molecular aggregates.

2 INTERNAL VARIABLES

We will now try to reduce the effect of the numerous variables that affect the physical properties of an emulsion to some measurable parameters. The main internal variables are composition of either phase and of the surface layer of the droplets, and the volume fraction and size (distribution) of the droplets. Table 1 gives a qualitative summary of the interrelations; it is to some extent an oversimplification. To see how the table may be used, consider Flocculation (number 5): which depends, among others, on variable 1 (interaction

Table 1 Main internal variables affecting properties of oil-in-water emulsions

Variable	Affected by					
	Variable number	Surface layer	Droplet size distribution	Volume fraction of oil	Compos. aqueous phase	Compos. oil phase
1. Interaction energy						
van der Waals attraction		x	x			
electrostatic repulsion		x	x		x	
steric interaction		x	x		x	
bridging		x	(x)	(x)	x	
2. Crystallization of oil		(x)	x			x
3. Crystallization of water			(x)	(x)	x	
4. Air-oil interactions	1, (2)	x	(x)	(x)	x	x
5. Flocculation	1, (3)		(x)	x	x	
6. Creaming	(2, 3, 4), 5	(x)	x	x	x	
7. Rheological properties	1, (2, 3, 4), 5		(x)	x	x	
8. Coalescence	1, (2, 3, 4, 5), 6	x	x	x	x	(x)
9. Disruption	(2, 3, 4)	x	x	x	(x)	x

energy between droplets) and on the composition of the aqueous phase. No cross is put under "surface layer", not because it has no effect on flocculation, on the contrary, but its primary action is via the interaction energy. The aqueous phase also affects the interaction energy, but it also influences flocculation via its viscosity. Numbers or crosses between brackets imply that the variable may have an effect, according to other conditions. Exceptional or very weak effects have been omitted. Most of the items mentioned in Table 1 will be discussed below.

3 INTERACTION ENERGY

The interaction energy G_{int} is the free energy needed (or gained if its sign is negative) to bring two droplets from an infinite distance apart to some close distance h; it is thus a function of h. G_{int} usually is the sum of three terms, originating from Van der Waals' attraction, electrostatic repulsion and steric repulsion. In many cases, G_{int} can now be calculated with some accuracy.[4,6,21]

The classical DLVO theory (Deryaguin, Landau, Verwey, Overbeek) considers the attraction generated by Van der Waals' forces (energy G_A) and the electrostatic repulsion caused by like charges on the particles (G_R). This is illustrated in Fig. 2, which gives an example of G_{int} as a function of h. In this case there is a maximum in G_{int} near B, a "primary" minimum near A and a shallow "secondary" minimum near C. Actually, at very small h, Born repulsion and hydration have to be taken into account as well.

In order to calculate G_{int} we have to know the radius of the particles (R), the Hamaker constant (A) (which depends on the materials in and between the particles[24]), the electric surface potential, ψ_0, of the particles (in principle to be derived from electrokinetic measurements) and the Debye-Hückel parameter κ. $1/\kappa$ is called the thickness of the electrical double layer and gives the distance over which the electric potential decays to $1/e$ of ψ_0; it depends on the total ionic strength I according to

$$1/\kappa \approx 0.3 I^{-0.5}$$

expressed in nm. All equations given only apply to aqueous solutions near room temperature and for $\kappa R \gg 1$ and $h \ll R$. In practice, ionic strength (thus κ), surface potential (influenced by pH and ionic composition) and particle size are important variables. Examples of the effect of these on the interaction energy are given in Fig. 3.

Another interaction energy (G_S), due to steric repulsion, should often be added. This type of repulsion results from molecular chains protruding from the surface of the droplets into the aqueous phase. Broadly speaking, the

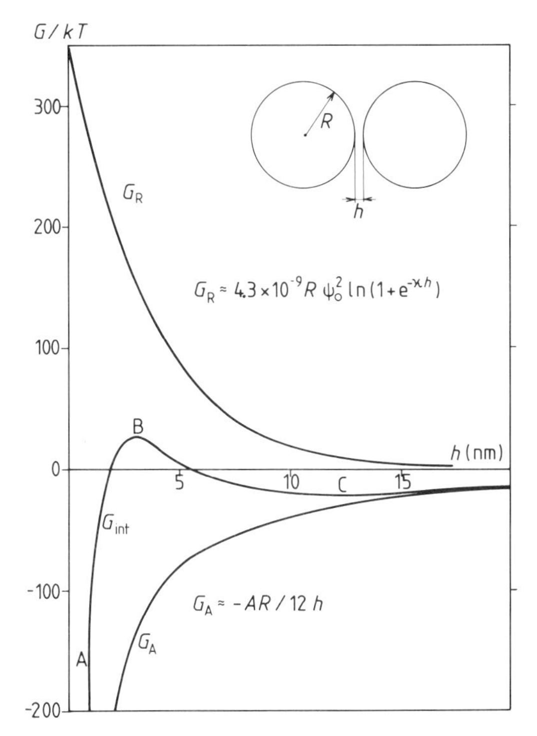

Fig. 2 Interaction energy (G) between two spheres of identical size at a distance h according to the DLVO theory. Calculated for $A = 10^{-20}$ J, $R = 2$ μm, $\psi_0 = 16$ mV and $\kappa = 3.10^9$ nm^{-1} ($I = 0.01$). The variables shown are G_R (due to electrostatic repulsion), G_A (due to Van der Waals attraction) and $G_{int} = G_R + G_A$

chains hinder one another when the droplets come close, thereby causing repulsion, at least when the chains are sufficiently hydrophilic. Calculation of G_S is difficult, but if sufficient macromolecular chains are present the repulsion can be very strong and be almost represented by a step function (see Fig. 4). It is seen that, again, a "secondary" minimum in the interaction energy may exist at a certain distance from the particle surface, and that such a minimum is deeper for larger particles.

Steric repulsion is usually caused by adsorbed macromolecules. The presence of such an adsorbed layer interferes also with the calculation of G_A (it affects the Hamaker constant) and of G_R (because it hinders determination of ψ_0, since electrokinetic measurements now yield the electric potential at some unknown distance of the particle surface). Nevertheless, the theory may give considerable understanding of trends when altering conditions.

The surface layers largely determine the interaction energy and thereby

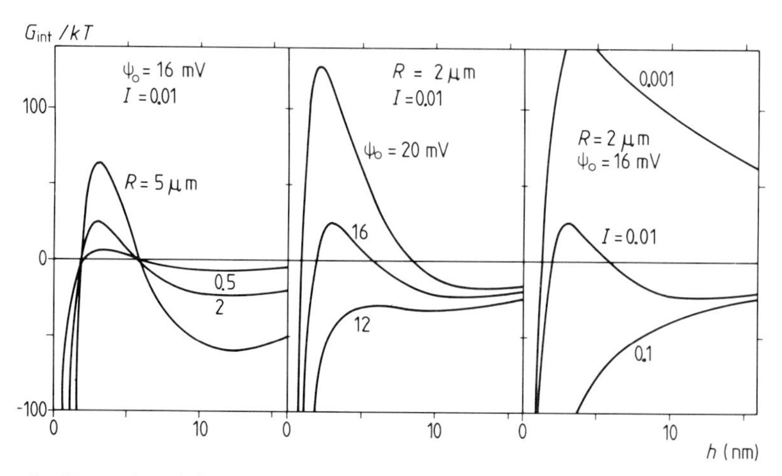

Fig. 3 Examples of the effect of some variables on the total interaction energy (G_{int}) between two identical spheres at a distance h, calculated from the DLVO theory for $A = 10^{-20}$ J

whether flocculation of the droplets can occur (see below), which in turn strongly affects rheological properties and creaming rate. The aqueous phase affects G_{int}, as pH, ionic strength and solvent quality influence electrostatic repulsion (via surface charge and its decrease with distance from the droplet surface) and steric repulsion (via the conformation of protruding mocular chains). Even the sign of G_{int} can reverse. In a food emulsion such properties

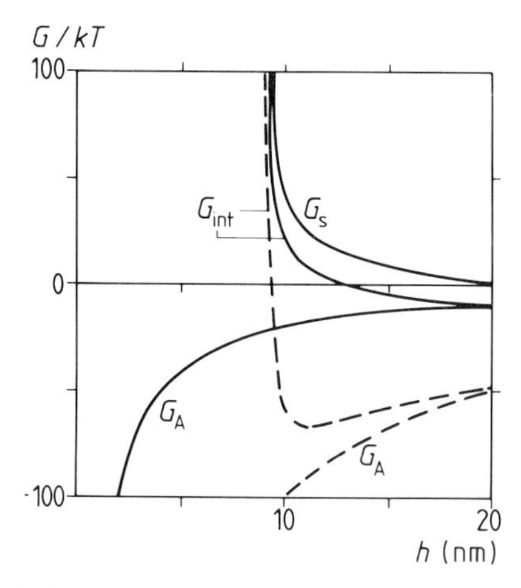

Fig. 4 Examples of the interaction energies (G) between spheres of identical radius R, due to steric repulsion (G_s) and Van der Waals attraction (G_A). $G_{int} = G_S + G_A$. $R = 1$ μm (——) and 5 μm (– – – –). Highly schematic

(e.g. pH) may alter during processing or storage. Moreover, bridging may occur, for instance between protein-covered droplets when the proteins undergo cross-linking reactions (as may happen during heating).

4 FLOCCULATION

The interaction energies as given in Figs 2–4 have been divided by kT, which is the average kinetic energy involved in an encounter between two molecules or two particles in Brownian motion. The maximum in G_{int} (e.g. near B in Fig. 2) can be interpreted as an energy barrier or activation energy G^* for flocculation "in the primary minimum", that is two particles attaining "zero" distance between them. The probability that this happens for any encounter between particles is roughly $\exp(-G^*/kT)$ and if $G^* > 25kT$, that probability is virtually zero. If two emulsion droplets overcome the energy barrier, and thus come into close contact, they will coalesce (see below). Two droplets may, however, become and stay flocculated "in the secondary minimum" (that is at some mutual distance, as near C in Fig. 2), if the minimum G_{int} is several times kT below zero. It thus follows that the larger the particles the greater the probability that they flocculate. Flocculation can also be caused by several types of bridging between particles or their adsorbed layers, and this may involve any of several, highly specific interactions.

One should also consider the rate of flocculation. If the rate of encounters is determined by Brownian motion, and if there is no energy barrier (as is the case for flocculation in the secondary minimum), it follows from Smoluchowski's theory[4] that the time needed to halve the number of particles (counting aggregates as one particle) is, in water at room temperature, roughly given by

$$t_{0.5} \approx 0.1d^3/\varphi$$

where t is in s, droplet diameter d is in μm and φ = volume fraction of droplets. Hence, flocculation usually proceeds very quickly. In practice, we often have a situation where particles are either flocculated or not and we may ignore the time needed to achieve it.

5 COALESCENCE

Coalescence of droplets into larger ones, which is the most characteristic physical change in emulsions, is very difficult to predict; there is no good theory. If two globules approach each other, a thin film of continuous phase remains between them; it is the breaking of this film that results in coalescence. The rate of coalescence largely follows first-order kinetics, while flocculation is typically a second-order reaction. Coalescence may be a rare phenomenon and still be noticeable. If, for example, only 1 in 10^5 of encounters between two droplets results in their coalescence, the emulsion may still show visual oil separation within a few days and thus be called "unstable".

It is often assumed that a main cause of increased coalescence stability is the lowering of the interfacial tension γ_{ow} by the surfactant, since the latter lowers interfacial free energy and thus minimizes the decrease in free energy upon coalescence as caused by the ensuing reduction in interfacial area. This is not true. To be sure, a surfactant is needed to make the emulsion, it is essential in preventing coalescence and a substance can only be a surfactant if it lowers γ; but a low γ is not the cause of the stability. On the contrary, a lower γ probably corresponds with a less stable system. The simplest stability theory as depicted in Fig. 5, gives as the activation free energy for coalescence $G^* \approx \gamma h^2$, where h is the thickness of the film between droplets; the theory proceeds from the assumption that the interface must be enlarged before contact between the droplets can be made. The theory is too simple, but serves to illustrate that a very close approach of the globules (say 2 mm) will always result in coalescence. Surfactants that keep the globules far apart without lowering γ too much are thus expected to impart good stability.

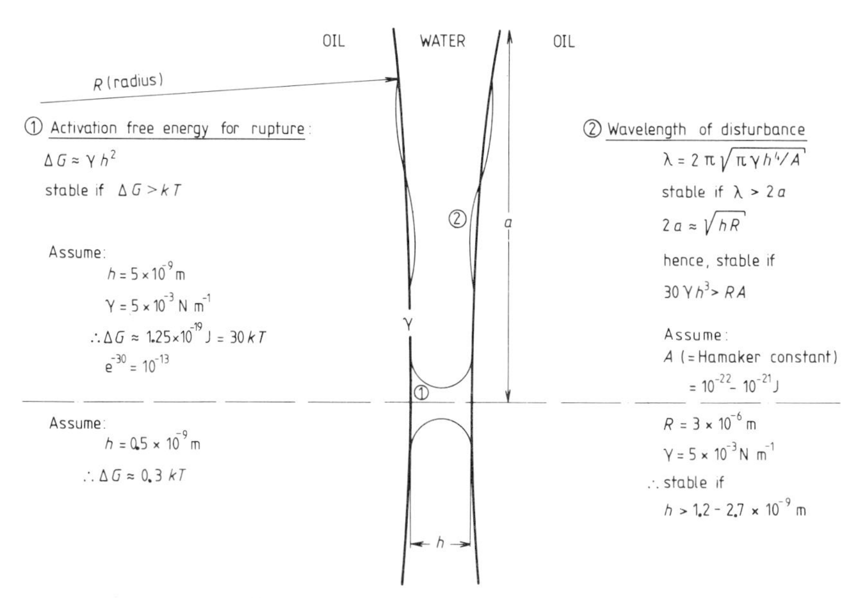

Fig. 5 Mechanisms of rupture of a thin liquid film between oil droplets. Highly schematic

More advanced theories calculate the stability of a thin film by considering the small wavelike deformations that will always occur due to thermal motion or external vibrations. If all such waves are damped, the film is stable; if the amplitude of any wave spontaneously increases, the film eventually ruptures.[8,9] Figure 5 gives an oversimplification of the theory, by considering the predominant wavelength of a disturbance. If this wavelength is longer than the diameter of the film, the latter will be stable. It can also be deduced that low γ and a thin film give less stability and, moreover, that a smaller film, hence smaller droplets, give greater stability. These results agree well with practical observations. If the theory predicts instability, it usually is correct, but if it predicts stability, it may or may not be so.

A full explanation of film instability is much more complicated. Dynamic surface properties should be considered. Theory predicts that a very-low concentration of surfactant is already sufficient to stop lateral movement of the interfaces bounding the film (and such movement would considerably enhance instability). But it is found in practice that other dynamic surface properties come into play; for instance, the shear viscosity of the o/w interface for the most part correlates well with emulsion stability. A somewhat intriguing problem is that several different surface properties can be measured which all correlate with stability, since they are interrelated, but it is *not* certain which is the operative factor determining film stability: rate of streaming of liquid out of the film; desorption of surfactant material from the interface; lateral compression of the interfacial layer? Nevertheless, useful predictions about the suitability of a surfactant (mixture) to stabilize emulsions often can be made.

From the above considerations it may be deduced that proteins are particularly suitable to make o/w emulsions. Most proteins are surface-active, soluble in the aqueous phase, cause considerable (steric) repulsion between droplets at a fairly large distance, and give a not very-low interfacial tension. The subject has recently been reviewed by Halling[7] and by Phillips.[18] Not all proteins, however, promote the same stability; for instance, caseinate is much more effective than gelatin. pH and ionic strength considerably affect emulsion stability, and not in the same way for all proteins. But a fair correlation between surface shear viscosity and the stabilizing power of the protein is usually found. Naturally, the thickness of the protein layer should be sufficiently high, say a surface excess of 2 mg m^{-2}. If the droplets are too large, coalescence nevertheless occurs (see Fig. 6).

Small amphiphilic molecules, often called emulsifiers, are also used in food emulsions, but most of them give less stable emulsions. They may serve several other purposes.[10] In small quantities added to proteins they usually make the emulsion far less stable (see Fig. 6), and it should be realized that many oils contain monoglycerides. Stable emulsions can often be made with

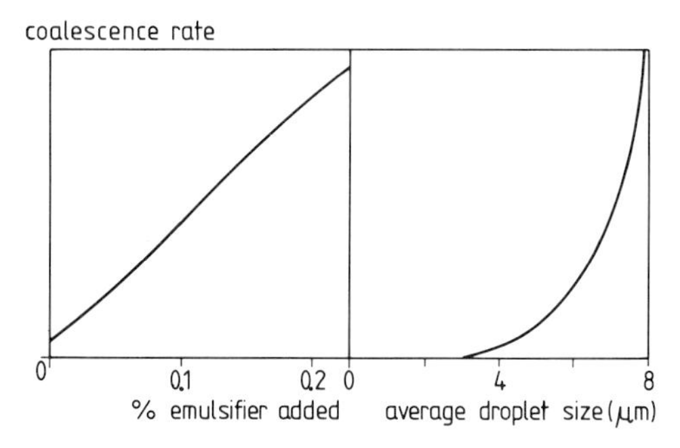

Fig. 6 Examples of the effect of added emulsifiers (e.g. polyoxyethylene sorbitan mono-oleate) and average droplet size on the rate of coalescence in protein-stabilized o/w emulsions. The illustrations are only meant to illustrate trends

emulsifiers if they form layers of a liquid crystalline phase around the droplets[6] but such structures require fairly high concentrations, since such layers are mostly >10 nm thick. Some progress has been made in the use of phospholipids forming liquid crystalline layers at fairly low concentrations.[19] But a suitable protein (mixture), if need be a modified protein, will fulfil almost any of these functions and is, of course, particularly suitable in foods. Emulsifiers may, however, serve to impart a controlled instability, for instance in whippable emulsions.[15]

6 PARTIAL COALESCENCE

The oil in the droplets of an o/w emulsion may (partially) crystallize (see Chapter 5 of this book). If by far the greater part of the oil in each droplet has crystallized, coalescence is no longer possible: the droplets behave like solid particles. But if a smaller proportion of the oil, say 10–50%, has crystallized, the droplets may show partial coalescence; they then aggregate into irregular granules that may be envisaged as rigid particles (rigid because they contain a network of flocculated crystals) kept together by "necks" of liquid oil. On heating the emulsion, the crystals melt and the granules turn into larger droplets.

If the droplets contain crystals, and if some of the crystals are located at the o/w boundary, the rate of coalescence is very much increased, maybe by as much as a factor of 10^6, particularly when the emulsion is agitated.[2] Examples of such results are in Fig. 7. The most probable explanation is that a crystal projecting into the aqueous phase pierces the thin film between closely approaching droplets, thereby triggering (partial) coalescence. Con-

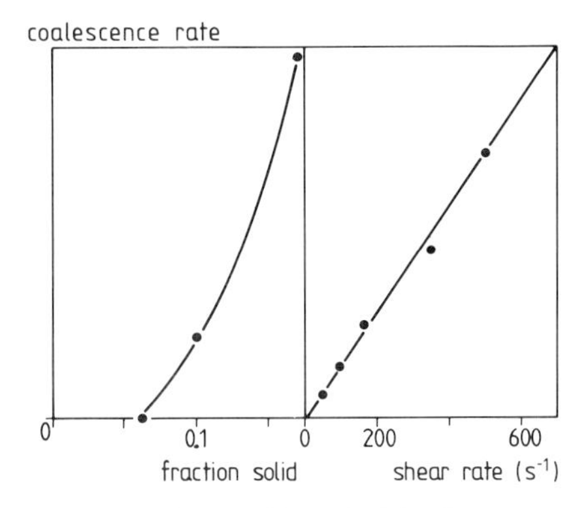

Fig. 7 Effect of the fraction of solid paraffin and of the shear rate on the coalescence rate of paraffin oil-in-water emulsions, stabilized by sodium lauryl sulphate, when subjected to Couette flow (after ref. 2)

sequently, if the interfacial tensions between oil, water and crystals, as influenced by surfactants, are such that the crystals project further, then the stability to coalescence should be less, as has been observed.[3]

Incidentally, if droplets are fully covered by solid particles that are predominantly projecting into the continuous phase, they are effectively protected against coalescence (though not against flocculation); this is called Pickering stabilization.[11] This is considered to be responsible for the resistance to coalescence of the water droplets in margarine, which are covered by fat crystals.

7 EXTERNAL VARIABLES

Besides composition and structure of the emulsion (the internal variables), external conditions may greatly affect emulsion stability.

Temperature affects the surface properties and viscosity of the continuous phase; generally, a higher temperature implies decreased stability to creaming and to coalescence. A notable exception is the effect of temperature on fat crystallization and hence on coalescence stability.

Creaming. If the emulsion droplets are allowed to cream, coalescence stability may dramatically decrease. In a cream layer, droplets are constantly close to each other, instead of occasionally and briefly as in the bulk. Moreover, droplets may deform each other, thereby enlarging and thinning the films between them. Consequently, the rate of coalescence will usually be greatly increased, particularly if the droplets are large.

The creaming rate is according to most textbooks found from the Stokes' equation

$$v = a\,\Delta\rho d^2/18\eta_c$$

which relates the velocity v of a submerged sphere relative to the surrounding liquid, to the acceleration a (mostly $a = g$), the difference in density $\Delta\rho$, the sphere diameter d and the (Newtonian) viscosity of the continuous phase η_c. This implies, for instance, that the rising rate of 1 μm oil droplet in water is some 5 mm/d. However, many conditions must be fulfilled for the Stokes' equation to hold,[27] such as an infinitesimally small volume fraction φ and particles not smaller than roughly 1 μm. Figure 8 gives examples of the effect of φ on v.

Nevertheless, the equation indicates the possibilities for diminishing creaming rate, such as decreasing d and increasing η_c. Prevention of creaming can be achieved by giving the continuous phase a yield stress, for instance by adding a suitable polysaccharide. The yield stress needed is of the order of $gd\,\Delta\rho$, which implies only some 10^{-2} Pa for 10 μm oil droplets in water. Consequently, a very weak gel suffices. A yield stress preventing creaming may also be formed when the droplets themselves flocculate into a continuous network of sufficient strength. If φ is low, however, flocculation

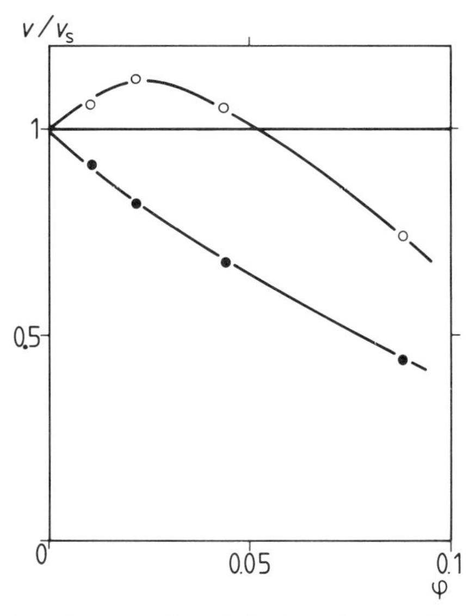

Fig. 8 Influence of the volume fraction of oil (φ) on the creaming rate (v), divided by the rate predicted according to the Stokes' equation (v_s) in gravity (●) and centrifugal creaming (○) of o/w emulsions with an average droplet diameter of approximately 1 μm (after ref. 27)

often leads to the formation of separate aggregates that cream much faster than the single droplets do.

Agitation of the liquid implies the addition of energy and therefore will increase the encounter frequency of droplets and the rate of flocculation. However, more importantly it may disrupt floccules and if agitation is very vigorous, it may even cause disruption of (large) droplets; such disruption cannot occur without the input of external (usually mechanical) energy. Coalescence is rarely influenced by agitation,[1] unless the droplets are partly solid[2] (see Fig. 7). In the latter case agitation promotes coalescence and the presence of solid fat also prevents the disruption of the larger aggregates formed, which would occur if the droplets were fully liquid. Consequently, most o/w emulsions are very unstable when agitated under conditions where part of the oil is crystallized.

Other factors. The beating in of air frequently causes the disruption of oil droplets, since oil tends to spread over the o/w interface; if the oil droplets are partially solid, coalescence (churning) occurs.[15] Coalescence of droplets may easily occur during freezing, evaporating or drying of an emulsion.

8 EMULSION FORMATION

We have seen that two variables are of paramount importance in emulsion stability:

—the droplet size: the larger it is, the higher creaming rate, coalescence rate and probability of flocculation;
—the surface layer, which largely determines whether flocculation can occur and greatly affects the rate of coalescence.

Both variables do not only depend on the overall composition of the emulsion (the ingredients used), but also on the way it is made. Emulsion formation has recently been reviewed.[25]

Droplet size. During emulsification large droplets are formed initially and these are subsequently disrupted into smaller ones. In order to disrupt a droplet of diameter d, the mechanical stress applied to it should overcome the Laplace pressure of the droplet, given by $4\gamma/d$, where γ is the interfacial tension between both phases. Hence, the smaller the droplet, the more difficult it is to disrupt it. The stress acting on the droplet can be a shearing stress, roughly given by $\eta_c\dot{D}$, where η_c is the viscosity of the continuous phase and $\dot{D}$ the deformation rate (for instance the shear rate) in liquid. Values of $\dot{D} > 10^4\ s^{-1}$ can be realized only with difficulty and in water ($\eta_c \approx 10^{-3}$ Pa s) stresses are thus at most 10 Pa, leading to droplets of some 1 mm (if $\gamma \approx 5\ mN\ m^{-1}$). Consequently, stress must be applied in some other way. It is usually derived from inertial forces in strongly turbulent flow; the essential parameter is the local energy density, i.e. the amount of energy dissipated per unit time and volume.

The *droplet size* attained therefore is highly dependent on the type of machine and on the intensity at which it operates. Figure 9 gives some examples. It is obvious that the high-pressure homogenizer is to be preferred for making o/w emulsions with small droplets. Another variable is the type of surfactant, primarily via the extent to which it lowers interfacial tension, but also in other (partly unknown) ways; it should be more soluble in the continuous (i.e. here the aqueous) phase. The average droplet size attained may vary by a factor 3 for different surfactants, macromolecular surfactants like proteins generally giving the larger droplets; this implies that more emulsifying energy must be used to obtain small droplets. Incidentally, droplet size is much more dependent on the type of surfactant if the emulsifying intensity is low, as with stirring or shaking. Several simple tests to assess the suitability of emulsifiers effectively are a measure of the droplet size

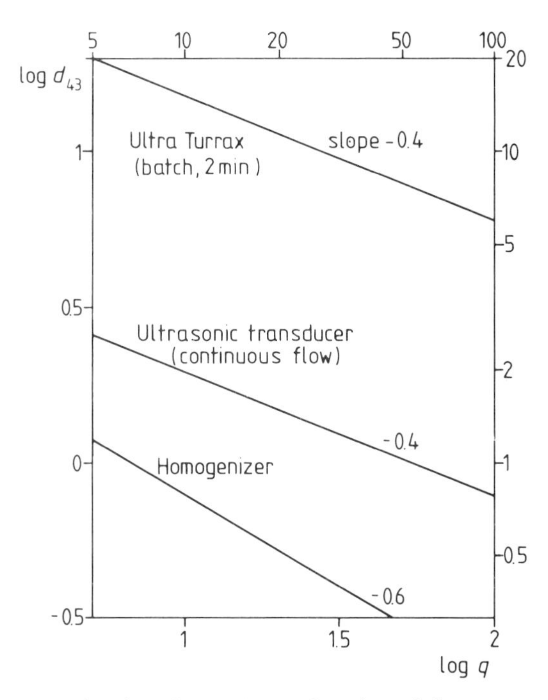

Fig. 9 Average droplet size (d_{43}, μm) as a function of the net energy input in the emulsion (q, MJ m^{-3}) for emulsification in various machines (from ref. 25 by courtesy of Marcel Dekker, Inc.)

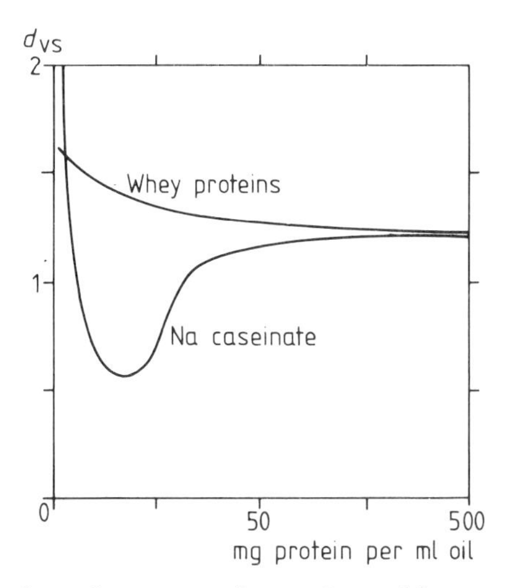

Fig. 10 The effect of protein concentration on the resulting average droplet size (d_{vs}, μm) when emulsifying oil in different protein solutions, other conditions being equal (from ref. 26)

obtained in such a way and may therefore correlate poorly with the stability of a properly produced emulsion.

The concentration of surfactant also affects droplet size. If there is insufficient to occupy the large interfacial area created, many droplets formed will immediately coalesce again. Coalescence of droplets during emulsification is a complicated phenomenon, primarily governed by factors that do not affect the coalescence of "finished" droplets. Figure 10 shows that the relation may be rather intricate; the upper curve is of the type more common for proteins (and other emulsifiers).

The *adsorption layer* will, for any surfactant, depend only on its concentration in the bulk phase after emulsification, provided that Gibbs equilibrium is reached. But a surfactant consisting of macromolecules, especially protein, usually adsorbs irreversibly (within the time considered) and the layer formed depends on conditions during emulsification.[15,22] It transpires that the thickness of the layer, here expressed as the surface excess Γ (in mg m^{-2}), is primarily governed by c/A, where c is the surfactant concentration (in the emulsion) and A the specific interfacial area (which, in turn depends on c). Examples are in Fig. 11; the linear relation between Γ and $\log(c/A)$ usually holds good, although there is no satisfactory explanation for it.

We will now exclusively consider proteins. A surface excess of some 1 mg m^{-2} corresponds to a monomolecular layer of unfolded peptide chains;

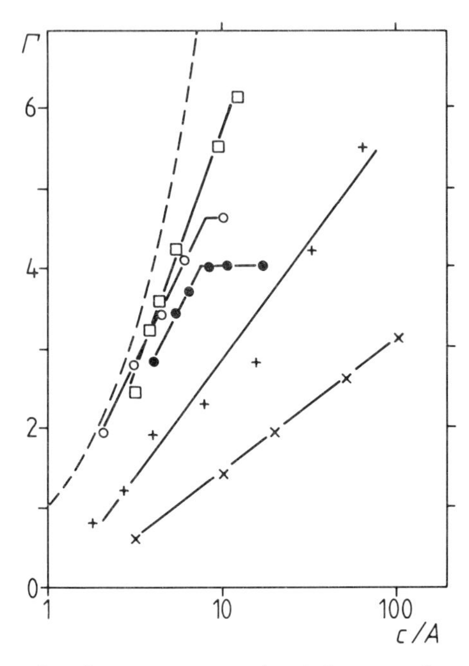

Fig. 11 The effect of surfactant concentration (c in mg m^{-3} emulsion) and specific interfacial area (A in m^2 m^{-3}) on the surface excess (Γ, in mg m^{-2}) obtained during emulsification with various macromolecular surfactants.

□ Partially esterified polymethacrylic acid; ○ and ●, two types of polyvinyl alcohol; + Na caseinate; × whey proteins. The dotted line would be found if all surfactant were adsorbed. After various sources

it can only be formed if c/A is small, so that adsorbed molecules have time and space to unfold at the o/w interface. Globular proteins adsorbed without much change in conformation give $\Gamma \approx 3$ mg m^{-2}. If $\Gamma > 5$ mg m^{-2}, then molecular aggregates have generally been adsorbed. It will be clear that Γ must depend on the type of protein (e.g. molecular weight, conformation) and on the conditions in the aqueous phase affecting protein conformation and aggregation (or solubility). Γ may, also, depend on the propensity of the protein to change its conformation.[18]

The surface excess is of considerable practical importance. First, it determines how much protein is needed to make an emulsion with some desired volume fraction and globule size. If the protein forms large aggregates (and therefore is poorly soluble), very much protein is needed and it may be said that such a protein is a poor emulsifier. If there is, as is common, a mixture of small and large species (e.g. separate molecules and aggregates), the larger ones are preferentially adsorbed during emulsification.[28]

Moreover, protein aggregates and large molecules may form bridges between droplets by becoming adsorbed onto two droplets simultaneously, if there is insufficient protein to cover the o/w interface; this causes aggregation and a marked increase in viscosity.[16] Finally, the surface excess is directly related to coalescence stability. If Γ is low (say <1 mg m^{-2}) the emulsion probably is unstable, especially at high temperatures. Addition of small-molecule emulsifiers usually causes a decrease in protein surface excess along with a decrease in stability.

References

1. Boekel, M. A. J. S. van and Walstra, P. (1981). *Coll. Surf.* **3**, 99–107.
2. Boekel, M. A. J. S. van and Walstra, P. (1981). *Coll. Surf.* **3**, 109–118.
3. Darling, D. G. (1982). *J. Dairy Res.* **49**, 695–712.
4. Dickinson, E. and Stainsby, G. E. (1982). Colloids in Foods, Applied Science Publ., London.
5. Farinato, R. S. and Rowell, R. L. (1983). Encyclopedia of Emulsion Technology, Vol. 1, P. Becher, ed., pp. 439–479, Dekker, New York.
6. Friberg, S. (1976). Food Emulsions, pp. 1–38, ed. S. Friberg, Dekker, New York.
7. Halling, P. J. (1981). *CRC Crit. Rev. Food Sci. Nutr.* **15**, 155–203.
8. Ivanov, I. B., Jain, R. K. and Somasundaran, P. (1979). Solution Chemistry of Surfactants, Vol. 2, ed. K. L. Mittal, pp. 140–167, Plenum, New York.
9. Jain, R. K., Ivanov, I. B., Maldarelli, C. and Ruckenstein, E. (1979). Dynamics and Instability of Fluid Interfaces, pp. 817–840, ed. T. S. Sørensen, Springer, Berlin (Lecture Notes in Physics **105**).
10. Krog, N. and Lauridsen, J. B. (1976). Food Emulsions, pp. 67–140, ed. S. Friberg, Dekker, New York.
11. Lucassen-Reynders, E. H. and van den Tempel, M. (1963). *J. Phys. Chem.* **67**, 731–734.
12. McNulty, P. B. and Karel, M. (1973). *J. Food Technol.* **8**, 309–331, 415–427.
13. Matsumoto, S. and Sherman, P. (1981). *J. Texture Stud.* **12**, 243–257.
14. Matsumoto, S., Ueda, Y., Kita, Y. and Yonezawa, D. (1978). *Agr. Biol. Chem.* **42**, 739–743.
15. Mulder, H. and Walstra, P. (1974). The Milk Fat Globule: Emulsion Science as Applied to Milk Products and Comparable Foods, C.A.B., Farnham Royal and Pudoc, Wageningen.
16. Ogden, L. V., Walstra, P. and Morris, H. A. (1976). *J. Dairy Sci.* **59**, 1727–1737.
17. Oortwijn, H. and Walstra, P. (1979). *Neth. Milk Dairy J.* **33**, 134–154.
18. Phillips, M. C. (1981). *Food Technol.* **35**, 50, 52, 54–57.
19. Rydhag, L. and Wilton, I. (1981). *J. Am. Oil Chem. Soc.* **58**, 830–837.
20. Sherman, P. (1983). Encyclopaedia of Emulsion Science, Vol. 1, pp. 405–437, ed. P. Becher, Dekker, New York.
21. Tadros, T. F. and Vincent, B. (1963). Encyclopedia of Emulsion Technology, Vol. 1, pp. 129–285, ed. P. Becher, Dekker, New York.
22. Tornberg, E. (1978). *J. Sci. Food Agr.* **29**, 867–879.

23. Verrips, C. T., Smid, D. and Kerkhof, A. (1980). *Eur. J. Appl. Microbiol. Biotechnol.* **10**, 73–85.
24. Visser, J. (1972). *Adv. Colloid Interface Sci.* **3**, 331–363.
25. Walstra, P. (1983). Encyclopedia of Emulsion Technology, Vol. 1, pp. 57–127, ed. P. Becher, Dekker, New York.
26. Walstra, P. and Jenness, R. (1984). Dairy Chemistry and Physics. Wiley, New York.
27. Walstra, P. and Oortwijn, H. (1975). *Neth. Milk Dairy J.* **29**, 263–278.
28. Walstra, P. and Oortwijn, H. (1982). *Neth. Milk Dairy J.* **36**, 103–113.

7 Kinetic Aspects of Food Emulsion Behaviour

D. F. DARLING

Unilever Research, Bedford

1 INTRODUCTION

The description "food emulsion" represents a broad characterization for a wide variety of foodstuffs which contain a dispersion of either a lipid or an aqueous phase, one or both of which may be either liquid or partially crystalline (e.g. salad dressings and icecream). With such a broad definition it is not surprising that the physical and chemical properties of food emulsions are diverse. There have been few texts which attempt to identify the generic

FOOD STRUCTURE AND BEHAVIOUR
ISBN 0-12-104230-8

issues in food emulsion science which describe the physicochemical properties, particularly those relating to stability.

Classical works on emulsion science (see for example refs 1, 38) consider many of the basic principles of emulsion stability, particularly the description of interaction energies between two approaching droplets. In the previous chapter by Walstra, the essential features of the stabilization mechanisms have been described. The destabilizing force is attributed to van der Waals' attraction whilst stabilization is related to either electrostatic or steric repulsion. Whilst a knowledge of these principles is essential in understanding the physical properties of colloidal dispersions they have only a general relevance where predictability is required. Emulsion stabilization has been treated on the grounds of thermodynamic arguments which, whilst representing the fundamental driving force for any physical or chemical change to occur, does little in predicting the rate of change. Unfortunately, as a food technologist, it is this predictability which is essential to any industrial research and development programme.

In food science, the important physical properties of emulsions are invariably manifested as a function of time. Whether it be the whipping of cream, the serum separation from a sauce or the shrinkage of icecream. They are time dependent phenomena and as such, are kinetic events.

One of the main reasons why thermodynamic arguments are imprecise and misleading is that a kinetic event invariably involves overcoming an energy barrier. It is the magnitude of the energy barrier that limits any rate of change. Unfortunately the absolute calculation of energy barriers is only possible in simple, well defined, model systems—far removed from the complex world of food emulsions.

The measurement of rate processes, whilst not describing the mechanism by which a change occurs does characterize the change in a relevant way. The conversion of milk into cream involves centrifugal separation of emulsion droplets, the efficiency of which can be described in terms of particle size, aggregation and particle density. The subsequent aeration of cream and conversion to butter involves adsorption of emulsion droplets to air cells, and droplet aggregation. The rate at which these processes occur defines the churning time and to some extent fat loss in the serum from the churned product. The consumption of butter involves melting of fat and release of water droplets. The perceived saltiness of butter relates to the coalescence rate of water droplets with saliva and the release of flavour components.

The above examples illustrate the significance of kinetic processes in describing the behaviour of food emulsions. Furthermore, by characterizing these kinetic processes a unique description of emulsion stability can be achieved which can be used to predict the physical behaviour of the emulsions under realistic circumstances.

The various physical properties of food emulsions are influenced by the stability, or rather instability of the dispersed phases. Whilst the effects of instability are exhibited in different ways for different foodstuffs, the underlying mechanisms by which these processes occur can be described in terms of three basic kinetic events—separation, adsorption and aggregation. Although instability is often undesirable (as in the case of serum separation) there are many instances when it is beneficial and indeed essential to food processing and properties (as in the conversion of milk into butter).

Whilst thermodynamics governs the origin of emulsion instability, the manifestation of loss of stability in any of its many forms is a kinetic process. The basic kinetic events and their interactions are summarized in Fig. 1. The aim of this chapter is to review the level of understanding in each of these areas, describe the relevant theoretical aspects and where possible illustrate with practical examples.

Figure 1 illustrates that aggregation can be classified into flocculation (where droplets retain their identity) and coalescence, where their identity is

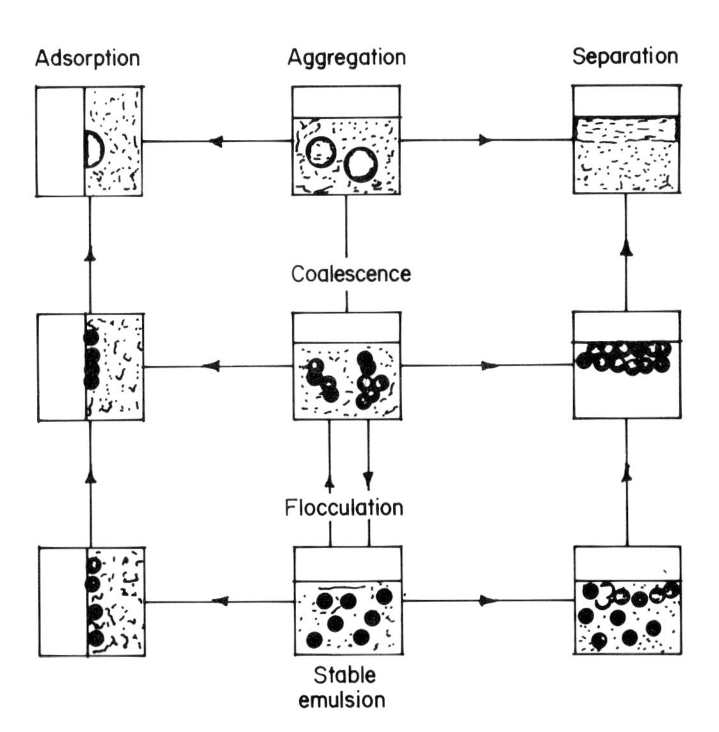

Fig. 1 Schematic illustration of the various kinetic processes involved with emulsion instability

lost. These definitions are consistent with classical descriptions of emulsion instability, but require some expansion. Flocculation in colloid science is usually used to describe aggregation of particles in a weak secondary minimum, while coagulation is reserved for aggregation in a primary energy minimum. However, in high phase volume dispersions, as is the case with many foods, the concept of a secondary energy minimum is difficult to define because the droplets are so close together. In addition, depending upon the degree of crystallization of the emulsion droplets, coagulation may or may not result in coalescence. For the purposes of this review therefore, flocculation is used to describe any aggregation process in which the droplets retain their identity, and coalescence for when they lose it.

The same processes can occur at interfaces or during separation. Thus the total picture of the kinetic processes involved in describing emulsion behaviour can become extremely complex. In food systems complex kinetic processes are the norm—the whipping of cream involves the adsorption and aggregation of fat droplets;[13] the creaming of milk involves aggregation and separation.[29]

Before continuing with a more detailed description it is worth noting that kinetic events are frequently reversible. In the context of food emulsions reversibility is particularly significant in the case of flocculation.

A consideration of the kinetics of instability phenomena provides a unique means of describing emulsion stability in practical systems. The rate constants associated with each instability process can be used as standard criteria for emulsion instability having both uniqueness and practical significance.

2 SEPARATION PROCESSES

Separation, commonly called creaming, is manifested by a partitioning of the disperse phase as a function of time. Separation usually only occurs when there is a finite density difference between the disperse and the continuous phase, the driving force is then either gravitational or centrifugal. However, occasionally in high phase volume systems, a network of emulsion droplets can be formed which exhibits syneresis, "squeezing out" the continuous phase. Such a process can sometimes be observed in salad cream, or homogenized dairy cream. The mechanisms of syneresis will not be covered here; suffice it to say that such processes do occur and can occasionally play an important role in food behaviour.

The simplest case of separation describes the velocity of an isolated, rigid, spherical particle moving through a homogeneous medium (Stoke's

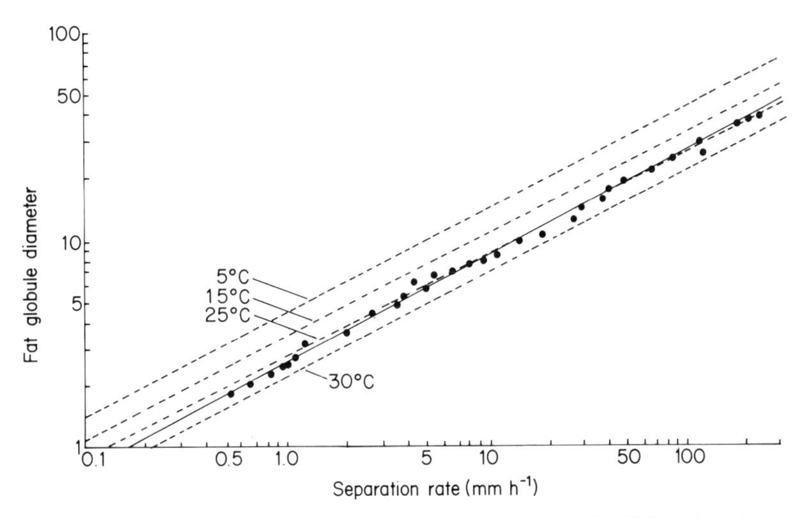

Fig. 2 Effect of fat globule diameter on separation rate of individual globules in milk at 25 °C. Dotted lines refer to Stokes' law predictions at different temperatures. (Reproduced with kind permission from [52])

equation):

$$V_s = \frac{a(\rho_d - \rho_c)d^2}{18\eta} \tag{1}$$

Where a is 9.81 m.s^{-2} for gravity creaming, and $r\omega^2$ for centrifugal separation (r, the radius of rotor arm and ω, the angular velocity), d is the particle diameter, η the medium viscosity and ρ_d and ρ_c the densities of the disperse and continuous phases respectively. Typically a 1 μm oil droplet ($\Delta\rho \sim 100$ kg m^3) in water ($\eta \sim 1$ mPaS) would move at about 5 mm per day.

Troy and Sharp[52] have demonstrated that Stokes' equation is obeyed for single isolated fat droplets in milk (Fig. 2) under very dilute conditions and in the absence of convection currents. These extreme conditions are rarely, if ever, found in food emulsions, and it is not surprising therefore that this equation is inadequate in practice. Some of the factors affecting the validity of Stokes' equation are described below.

A Deviations from Stokes' equation

1 FACTORS ENHANCING SEPARATION RATE

Particle aggregation leads to an increased effective hydrodynamic diameter and hence enhanced separation rate. There are many types of particle interaction which can lead to aggregation including flocculation in a primary or secondary energy minimum, polymer bridging flocculation, and coalescence. These have been discussed in more detail in the preceding chapter. Particle-particle interactions in foods are common-place and can lead to enhanced separation rates.

The derivation of Stokes' equation assumes that there is no slippage at the interface between particle and continuous phase. This is only strictly true for highly viscous/solid particles, although the presence of a surfactant which imparts viscoelasticity to the interface often means that the droplet can be treated as a hard sphere.

2 FACTORS RETARDING SEPARATION RATE

The most serious discrepancy between theory and experiment is the presence of neighbouring particles. As the disperse phase volume increases the particles cannot be treated as isolated species. The flux of liquid in the opposite direction retards the particle velocity. Particle-particle collisions also retard creaming rate; this is particularly true in polydisperse systems where small particles (rising slowly) collide with faster moving larger particles. No satisfactory theory has been developed for these interaction effects (so called hindered settling) although some progress has been made in the chemical engineering field. This will be treated in a subsequent section.

Dickenson[18] has pointed out that charged particles do not approach as closely as pairs of hard spheres, thus the effective phase volume is increased and the hydrodynamic effects discussed above become more pronounced. Presumably the same phenomenon should occur with adsorbed macromolecules although this has not been treated theoretically.

The viscosity of the continuous phase in food emulsions is rarely Newtonian and hence the viscous drag on the particle will itself depend upon the separation rate. This can, of course, be calculated from the measured creaming rate, but cannot be predicted analytically since one depends on the other. There are many polysaccharides which exhibit pseudoplastic behaviour such that at low shear rates they exhibit high viscosities and hence very low separation rates. Xanthan gum is well known in the food industry, on this basis, for its ability to stabilize dispersions against separation.

Stokes' equation assumes all particles have a constant density. Whilst this may be reasonable for large particles, when emulsion droplets approach 1 μm or less the presence of a protein membrane can increase the droplet density, particularly if the protein layer is thick as is the case when casein micelles are adsorbed (0.02–0.1 μm). This effect explains the observation by Fox *et al.*[19] on the homogenization induced complex between fat and milk proteins. The newly formed membrane as a result of homogenization results in the smaller droplets ($\lesssim 0.5$ µm) having an effective specific gravity greater than unity. These droplets therefore sediment in a gravitational field rather than cream.

Brownian motion of droplets may become significant as the particle size diminishes ($\lesssim 1.0$ µm). In fact oil droplets of a diameter 0.1 µm will remain in suspension almost indefinitely.

B Hindered separation

In most practical food emulsions, the concept of non-interacting particles is almost redundant because of the relatively high concentrations of disperse phase. As the disperse phase volume increases, the particles get closer together and begin to interfere with each other. The effect of this interaction has not been rigorously treated in terms of separation, but a variety of semi-empirical approaches have been developed from a chemical engineering standpoint.[50]

Three general approaches have been used to describe the process of hindered separation:

(1) Using an empirical correction term for Stokes' law as typified by the Richardson and Zaki[34] equation.

$$V_h = V_s^{4.65} \tag{2}$$

(2) Representing the continuous phase as a fluid with properties defined by the complete dispersion, thus the viscosity as experienced by one particle is influenced by the presence of others. Brinkman's theory[7-9] applied to the Einstein viscosity equation is the best example of this approach and yields:

$$V_h = V_s e^{4.5} \tag{3}$$

(3) In relatively high phase volume dispersions ($\gtrsim 0.5$) it is possible to consider separation as flow of liquid through a packed bed of particles. In this case the Kozeny-Carmen equation has been used[24,48]

$$V_h = V_s \frac{e^3}{10(l - e)} \tag{4}$$

Numerically all of these approaches produce similar results and can be summarized by the general expression

$$V_h = V_s e^2 \mathrm{f}(e) \tag{5}$$

where $\mathrm{f}(e)$ takes on slightly different values depending upon the approach adopted.

Most of these expressions are either empirical or at best based upon phenomenological approaches. Thus they may fit the observed data but generally only do so over a limited phase volume range. Davies *et al.*[16,17] have developed a generalized approach for hindered separation which, in brief, states:

$$V_h = V_s e^n \tag{6}$$

where n depends upon e according to:

$$n = \frac{e_l}{l - e_l} \tag{7}$$

and e_l represents the value of e when $V_h(l - e)$ is a maximum. Methods for calculating e_l (and hence n) are given by Davies *et al.*[17] and have been used to describe the settling of a wide range of dispersions. It is worth noting, however, that even in this generalized approach n only approaches infinity, when e approaches zero. In practice, separation will be infinitely slow long before $e = 0$. In fact, if the particles are rigid spheres then the maximum phase volume for random close packing $(1 - e)$ is $\sim$0.6 to 0.65 and under these circumstances n should be infinite for equation 6 to apply. There remains, therefore, some question about the value of this generalized approach for very concentrated dispersions.

Despite the attention given to separation processes in the chemical engineering field, little application has been made to foods. Walstra and Oortwijn[58] have published one of the few systematic studies of the application of hindered separation. In their study they investigated the effect of globule size and concentration on creaming in pasteurized milk and showed that the value of the exponent (n) was 8.6 which for relatively dilute emulsions ($<10\%$) is considerably higher than the empirical equations would predict. These observations by Walstra and Oortwijn illustrate that gravity creaming rates differ substantially from those predicted by Stokes' equation. For centrifugal creaming the deviations were not so large. The differences could be attributed in part to Brownian motion having a greater retarding effect in the case of gravity creaming.

Apart from work on milk, little published information exists in connection with separation processes. This is somewhat surprising considering the various situations in which separation may occur; for example, oil flavours in fruit drinks, creaming of coffee whiteners, serum separation in dressings and sauces, separation in premix tanks for various food processes, etc.

C Centrifugal separation

Centrifugation is used as a means of concentration either for subsequent manufacturing purposes, as in the preparation of cream from milk, or as a means of accelerated testing of emulsion stability.[40] Many of the aspects already covered are equally applicable to centrifugal separation and will not be repeated. Suffice it to say that the same limitations hold except that Brownian motion effects are usually negligible.

As a means of accelerated stability testing, centrifugation should be treated with caution. If the theories of hindered separation are used to predict gravity creaming on the basis of centrifugally enhancing the separation rate, it is implicitly assumed that all parameters in equation 5 are independent of acceleration (a) in this case ($r\omega^2$). This is rarely the case and is admirably demonstrated by some work of Vold and Groot.[55,56] They described the application of ultracentrifugation as a means of determining emulsion stability. In their studies, using paraffin oil emulsions stabilized with sodium dodecyl sulphate, they demonstrated a linear relationship between oil separation and centrifugation time (Fig. 3) the slope being an indicator of rate of separation and possibly emulsion stability.

Figure 3a demonstrates that for an emulsion containing 0.2% sodium dodecyl sulphate (SDS) and 50% paraffin oil the rate of oil separation decreases as the disperse phase volume increases, consistent with the theories of hindered settling. If the data were taken in isolation one might conclude that centrifugation would be a means of accelerating a storage test with respect to separation. However, the same authors showed that as the concentration of SDS was increased to 0.4% the effect of oil dispersed phase volume was completely reversed (Fig. 3b). Namely, the higher the phase volume of oil the faster the rate of oil separation.

The apparent inconsistency of these observations relates to the fact that oil separation involves two rate processes, droplet separation and droplet coalescence. The higher SDS concentrations produce more stable emulsions such that the rate of oil separation is controlled by the rate of coalescence and not by the rate of separation. As the disperse phase volume increases so the hydrostatic pressure on the droplets close to the surface of the centrifuge tube increases, thus rates of coalescence and oil separation also increase. With the less stable emulsion, containing 0.2% SDS, coalescence is no longer the rate determining process. Separation rate becomes a contributing factor in addition to coalescence. It can readily be shown that both rate processes are operative by determining the hindered settling exponent (n). Hindered separation experiments invariably result in values for the exponent >2.[50] The exponent calculated from the data in Fig. 3 is 0.5 implying that very little hindrance is occurring (unlikely in such concentrated emulsions) or that an additional rate process is reducing the droplet concentration (such as coalescence into the oil layer).

The above illustration demonstrates the importance of identifying the rate determining process and rigorously checking that the same process occurs

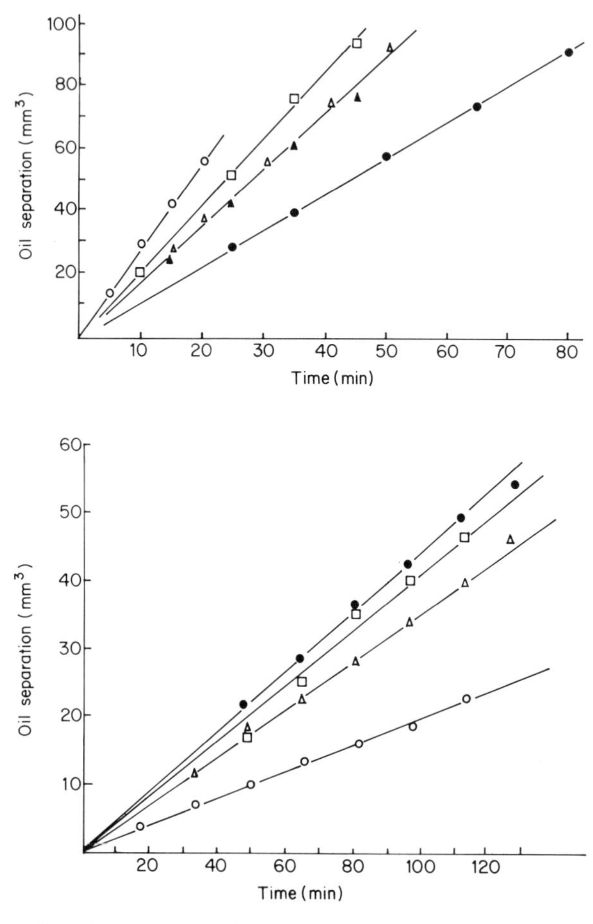

Fig. 3 Centrifugal separation of paraffin oil/water emulsions as a function of oil phase volume. (a) Emulsion prepared with 0.2% SDS. Oil Phase Volume. ○ 20.6%. □ 35.6%. △ 49.0%. ▲ 45.7%. ● 65.1%. (b) Emulsion prepared with 0.4% SDS. Oil Phase Volume. ○ 20.3%. △ 35.5%. □ 51.8%. ● 59.1%. (Reproduced with kind permission from [55,56])

when test parameters are varied. The work of Smith *et al.*[43] and Smith and Dainiki[44] on the stability of model o/w food emulsions, illustrates this necessity for caution. Their general conclusion, based upon gravity creaming experiments, was that emulsifiers increase the stability of protein stabilized emulsions. Unfortunately, under the conditions described it is likely that droplet flocculation by protein bridging occurred which would enhance creaming and hence decrease apparent stability. The two rate processes (flocculation and creaming) were not separated and the results therefore are inconclusive. The presence of emulsifiers will lead to protein displacement thus reducing flocculation and creaming. However, the membrane thickness will be reduced and coalescence could be enhanced (see Section 4.2).

Centrifugation can be used as a means of accelerated storage testing, but only under certain conditions. If stability refers to separation only, then the hindered separation equations can be applied provided due allowance is made for the effect of centrifugation on the other variables involved, e.g. viscosity. In particular, separation rate should be linearly dependent on centrifugal force. If not, all predictions should be treated with extreme caution.

3 ADSORPTION PROCESSES

Adsorption processes in colloid and interface science usually refer to the adsorption of surfactants to interfaces in the context of stabilization. In the previous chapter on the principles of emulsion science Walstra outlines some of these interfacial processes and the role they play in the formation, stabilization and destabilization of emulsions. This section deals with particle adsorption to interfaces and treats it as a kinetic event involving a variety of possible rate controlling mechanisms.

The practical significance and technological importance of adsorption processes in food emulsions is often overlooked. The aeration of cream, either for whipping or churning into butter, relies on the adsorption of fat droplets to air bubbles.[15] Similarly the textural properties of icecream are also highly dependent upon the degree of fat destabilization that has occurred during freezing and aeration.[29] The aeration of cake batters prior to baking can also depend upon similar adsorption processes.[35] In certain cases fat encapsulation of air is critical for the baking process to proceed correctly. Adsorption processes are not always beneficial; during processing of food emulsions, fat deposits can accumulate on pipework surfaces, on stirrer blades, or on the mechanical parts of a pump. During frying of moisture containing foods, such as meat, potatoes, pastry etc., the adsorption of hydrophilic particles (w/o emulsions) onto the metal surfaces results in the product adhering to the frying surface and subsequent burning.

Unfortunately, there have been few quantitative studies on adsorption phenomena as a kinetic process relevant to food emulsions. Most of the work in this area has considered adsorption from thermodynamic arguments in terms of reducing the total interfacial free energy of the system. In this context, the Young-Dupré equation is frequently used to describe the equilibrium position of an adsorbed droplet, namely:

$$\gamma_{12} = \gamma_{13} + \gamma_{23} \cos \theta \tag{8}$$

where the interfacial energies (γ) and cos θ are identified in Fig. 4. The work of adhesion, or the driving force (W) is then given by the energy required to detach the droplet from the surface, i.e.

$$W = \gamma_{23} + \gamma_{12} - \gamma_{13} \tag{9}$$

which combining with equation 8 leads to:

$$W = \gamma_{23}(l + \cos \theta) \tag{10}$$

If θ is 180° then droplets will spread and one can define a spreading coefficient as:

$$S = \gamma_{12} - (\gamma_{13} + \gamma_{23}) \tag{11}$$

which describes the driving force for an adsorbed droplet to spread across a surface.

In contrast to the above thermodynamic considerations, the kinetics of adsorption are poorly understood. The process involves a sequence of events

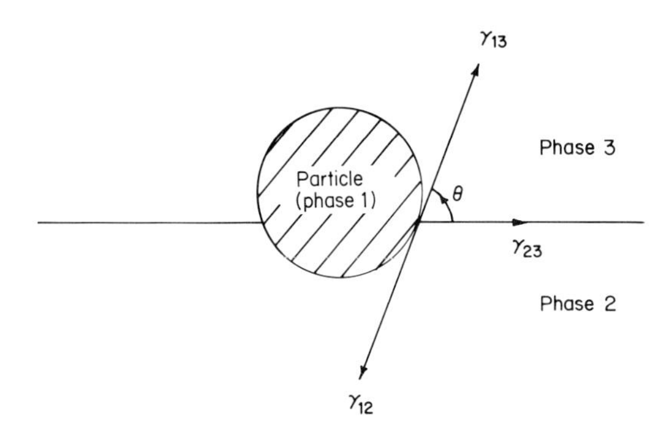

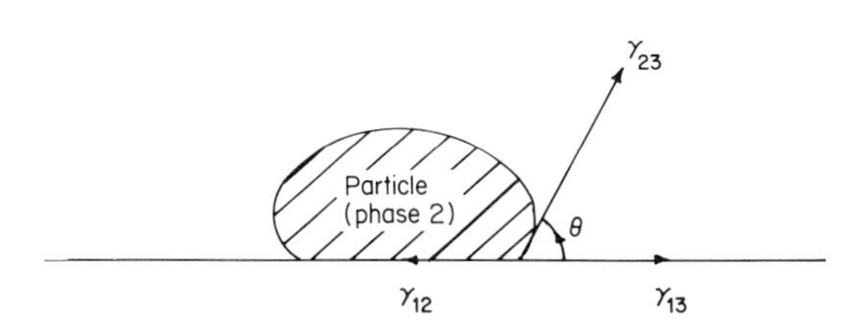

Fig. 4 Equilibrium position of a droplet adsorbed at an interface. (a) Solid droplet at a fluid/fluid interface. (b) Fluid droplet at a solid/fluid interface

as depicted in Fig. 5. The droplet approaches an interface, liquid begins to drain from the intervening film and the film thins. If the droplet is to absorb irreversibly the film must spontaneously rupture, and, having ruptured, the droplet must spread, eventually reaching some equilibrium conformation described by the balance of interfacial forces (Young-Dupré equation).

Within this complex picture there are many potential rate controlling stages which may dictate whether or not adsorption occurs:

1 Diffusion and convection processes will control the rate at which droplets come in contact with the surface. These are essentially the same as those discussed in a subsequent section on aggregation phenomena and will not be dealt with here. However, in practice with disperse phase volumes $>1\%$ and under agitated conditions, transport to the interface is unlikely to be rate determining.

2 Having approached the interface the rate of film drainage will be a function of viscosity, droplet size and Marangoni effects (see preceding chapter). Charles and Mason[10] describe the film thinning process between a droplet and an interface under gravitational forces by the expression:

$$t_{12} = \eta\left(\frac{\Delta\rho g d^5}{\gamma^2}\right)\left(\frac{1}{h_2^2} - \frac{1}{h_1^2}\right) \tag{12}$$

where t is the time taken for the film to thin from thickness h_1 to h_2. Mar and Mason[27] have extended this approach to consider a 3-phase system where the droplet is of a different composition from the interface. Under these

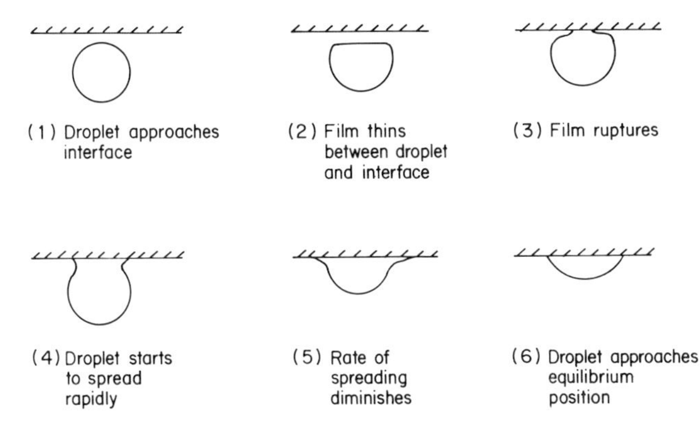

Fig. 5 Schematic illustration of the various processes involved in the adsorption of a droplet to an interface. (1) Droplet approaches interface. (2) Film thins between droplet and interface. (3) Film ruptures. (4) Droplet starts to spread rapidly. (5) Rate of spreading diminishes. (6) Droplet approaches equilibrium position

conditions the interfacial tension term becomes a harmonic mean of the three respective interfaces. The derivation of this expression assumes that the droplet distorts to give a planar interfacial film. Lang and Wilke[26] derived an alternative expression assuming the interface, not the droplet, deformed thus the film was spherical rather than planar. Under these conditions the result is:

$$t_{12} = \frac{3\eta}{4}\left(\frac{d^3}{\gamma}\right)J(p)\left(\frac{1}{h_2^2} - \frac{1}{h_1^2}\right) \tag{13}$$

where $J(p)$ is a trigonometric function describing the degree of penetration of the droplet into the surface. p is the angle subtended by the centre of the drop and the surface at the point of contact ($p = 90°$ for a hemispherical film and $0°$ when no penetration occurs).

As can be seen from these two examples, there are some common features notably the inverse square dependence of film thickness. Furthermore, both illustrate the strong dependence of droplet diameter. It is envisaged that the rate of adsorption of large droplets could be determined by film thinning but not with small droplets. It is worth noting that none of the approaches described include Marangoni effects which will tend to reduce the rate of film drainage.

3 Arguably the most critical stage is when the intervening film between droplet and interface ruptures. The mechanism of rupture is still uncertain although local fluctuations in film thickness brought about by environmental conditions are thought to be the driving force for overcoming the potential energy barrier (see preceding chapter by Walstra). As such, any irregularities in the surface will promote film rupture which supports the view that fat crystals in emulsion droplets have a destabilizing effect on stability.[4]

The kinetics of rupturing can be described in terms of a stochastic process. It is essentially a random event, governed by the laws of probability, that at some instant in time and some point in the film there is sufficient local energy to spontaneously break the intervening film. This will be considered in a little more detail later.

If rupture occurs before the film has thinned to some equilibrium value, then the rupture process will not be rate determining. If, however, the equilibrium film thickness is reached long before rupture occurs then the stochastic process becomes the determining factor. In food emulsions both situations can occur. For example, in premix tanks where relatively coarse unstable emulsions are frequently prepared prior to subsequent processing, large droplets come into contact with the stirrer blades, the container wall etc. and adsorption can be controlled by the rate of film drainage. Alternatively, in a stable food emulsion such as cream the oil droplets are small (~ 1 μm) with a thick protective protein layer and adsorption to surfaces are more likely to be goverened by film rupture.[14]

4 Having ruptured the film, the droplet, if sufficiently fluid, will spread until it reaches its equilibrium position. Under these circumstances the droplet viscosity and the respective interfacial energies will determine the rate at which spreading occurs.

Mar and Mason[27] derived an expression for the rate of spreading of a droplet at a planar interface. Accordingly they derived an expression which described the spreading rate:

$$-\frac{dH}{dt} \propto \left(\frac{\eta}{S}\right)^{0.5}\left(\frac{V^{0.5}}{t^{1.5}}\right) \tag{14}$$

where H is the average thickness of the spreading drop, V is droplet volume and S is the spreading coefficient (equation 11).

Using high speed cinematography they measured the spreading rate as a function of various parameters and found

$$-\frac{dH}{dt} \propto \left(\frac{\eta}{S}\right)^{0.53}\left(\frac{V^{0.6}}{t^{2.1}}\right) \tag{15}$$

Considering the complexity of the problem and many experimental difficulties this is in surprisingly good agreement with the theory.

Typical spreading rates for oil drops were found to be 10–100 cm/s. for 0.1 ml droplets. Thus spreading is complete within about 1 s. If, for comparative purposes only, one extrapolates back to very small emulsion droplets (say ~1 μm diameter) then the spreading rates would be very slow typically 10^{-5} to 10^{-4} cm/s. However, since the droplet is small the spreading distance is also small (~1 μm) and the process would still take about 1 s. To a first approximation therefore the time taken for a droplet to spread is almost independent of droplet size.

For small droplets the timescale is relatively long. For viscous droplets (e.g. those containing crystalline triglycerides) the rate of spreading is also so slow that other droplets will probably adsorb to the interface before the equilibrium position is reached. Alternatively, in many practical circumstances the emulsion is being agitated and desorption may occur before the droplet has reached its equilibrium conformation.

A Experimental studies on adsorption

One of the most critical stages in any adsorption process is the rupturing of the thin film between the droplet and the interface to which it is adsorbing. As already stated, this is considered to be a stochastic process and independent of droplet concentration.

Previous work on particle adsorption processes have been performed on the coalescence of single droplets with a macroscopic interface to investigate the mechanisms of droplet-droplet coalescence. Whilst the strategy for the latter approach is questionable it is instructive to consider the results of some of these studies as being pertinent to adsorption phenomena.

Experimentally, the lifetime of individual drops is monitored and the proportion of droplets surviving as a function of time is determined to yield a lifetime distribution curve (Fig. 6). These curves all have a common format, namely an induction period in which all droplets survive and a decay curve where droplets are gradually lost.

The induction period appears to relate to film drainage, which is described by Cockbain and Roberts[11] as the characteristic time required for the film to thin to some equilibrium thickness. The gradual decay in life time which follows this induction period can then be described by an equation of the form:

$$\ln N/N_0 = K(t - t_D)^n \tag{16}$$

where N/N_0 is the relative number of droplets remaining after time t. K is the rate constant. The exponent (n) is usually found to approach unity when film drainage to an equilibrium position occurs long before film rupture. However, if the droplets are relatively unstable such that rupture occurs during film drainage, then Gillespie and Rideal[20] have reasoned that $n = 1.5$.

The absolute value of n and its significance appears to be questionable. The data of Gillespie and Rideal clearly demonstrate that for benzene or paraffin oil droplets in water, N/N_0 is linear with respect to $t^{3/2}$. However, Silber and Mizrahi[42] studying the coalescence of orange oil droplets at an oil coated interface, used $n = \frac{3}{2}$ or 1 dependent upon the stabilizer being used (Figs 7 and 8).

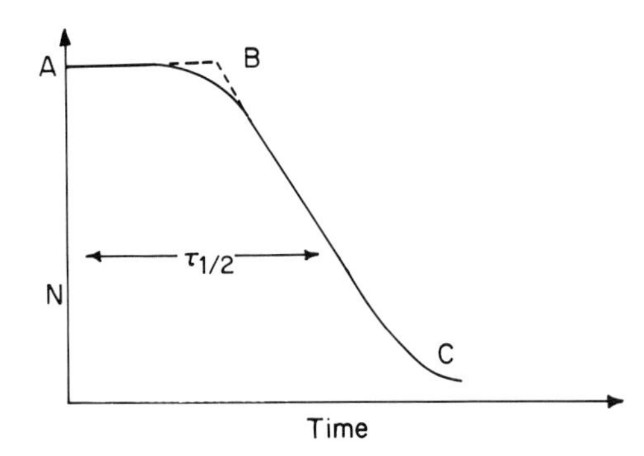

Fig. 6 Typical distribution curve for the lifetime of drops surviving at an interface. $\tau_{1/2}$, characteristic half life of drops. A–B, time required for film drainage. B–C, represents the process of film rupture. N, number of drops remaining after time (t)

Silber and Mizrahi used a selection of polysaccharide based stabilizers, all of which have some effectiveness in stabilizing emulsions. With propylene glycol alginate (Manucol Ester M) and carboxymethyl cellulose the data best fitted the equation with $n = 1.5$ while for gum arabic the data were best fitted for $n = 1$.

One of the biggest variables in any droplet lifetime experiment relates to the lifetime of the drop prior to bringing it in contact with the film. With

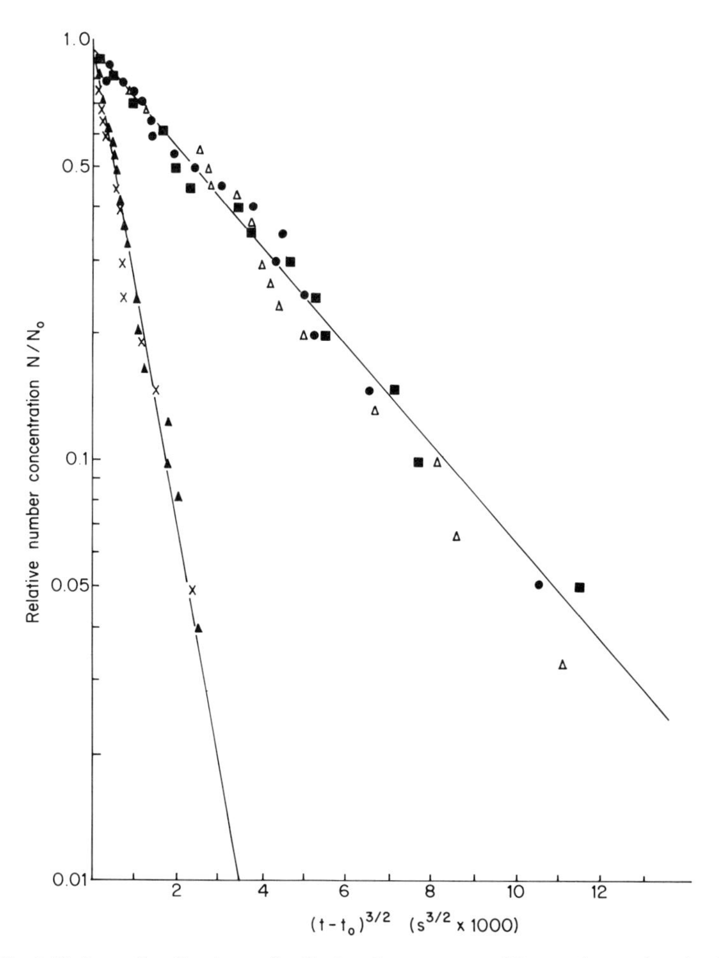

Fig. 7 Lifetime distribution of oil droplets at an oil/propylene glycol alginate solution interface. Data presented in terms of $(t - t_0)^{3/2}$ according to Gillespie and Rideal. Interface aged up to $1\frac{1}{2}$ h. Droplets aged up to 2 min. Gum concentration % w/v. ● 0.2. ■ 0.1. △ 0.05. ▲ 0.001. × 0.0001. (Reproduced with kind permission from [42])

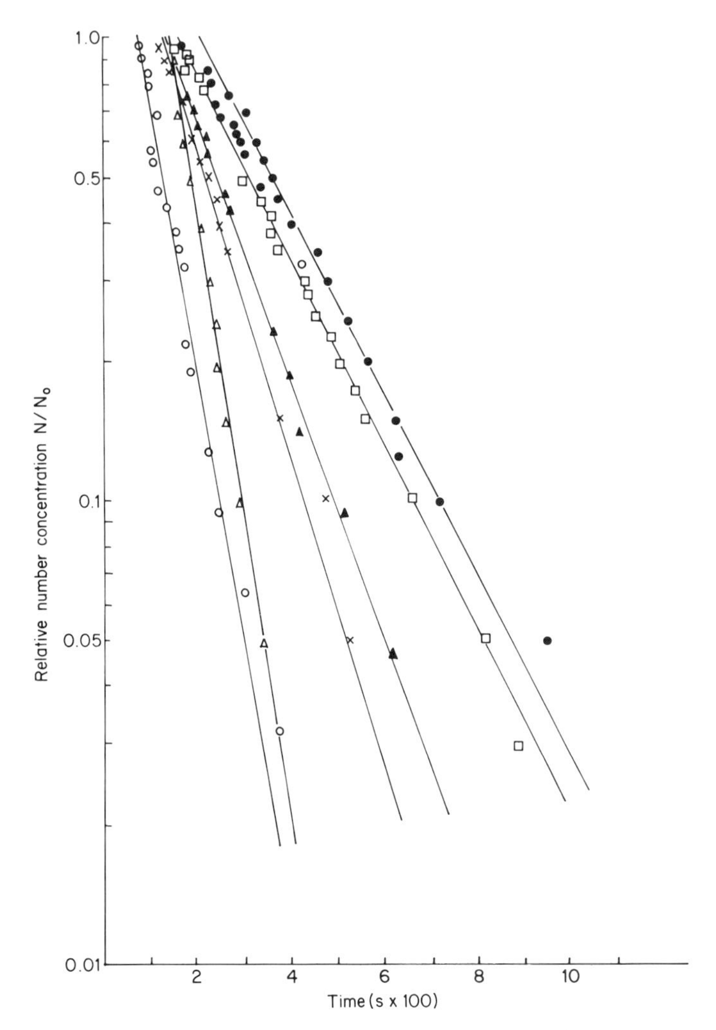

Fig. 8 Lifetime distribution of oil droplets at an oil/gum arabic solution interface. Data represented as a 1st order kinetic plot. Interface aged for up to 2 h. Droplets aged for 10 min. Gum concentration % w/v. ● 0.2. □ 0.1. △ 0.05. ○ 0.01. ▲ 0.001. × 0.0001. (Reproduced with kind permission from [42])

macromolecular stabilizers the molecular organization at the interface continues to change for several hours after droplet formation. This is reflected in the data obtained by Silber and Mizrahi where droplets left to equilibrate for 2 to 3 hours were an order of magnitude more stable than droplets equilibrated for shorter times.

A key parameter influencing film rupture is the equilibrium film thickness. Intuitively the thicker the film the lower the probability that surface irregularities will lead to rupture. Biswas and Haydon[3] demonstrated this with protein and surfactant stabilizers. In particular they demonstrated the ability of small molecule surfactants to displace proteins from interfaces and lead to a decrease in emulsion stability. The coalescence rate of bovine serum albumin (BSA) stabilized droplets increased with addition of an anionic detergent. The displacement of protein reduces film thickness and hence leads to a greater probability of film rupture. Surface rheology will also have a role to play, although few causative relationships with adsorption have been found.

An interesting demonstration of the effect of film thickness in a practical food emulsion is in the whipping of homogenized dairy cream. The whipping process involves the adsorption of fat droplets to air cells, gradually building up an encapsulated structure. The end point of the whipping process coincides with the maximum amount of fat adsorbed onto air cells.

Darling and Butcher[15] demonstrated that the interface of emulsion droplets in homogenized creams is composed primarily of casein micelles and their subunits. These protein particles are about 0.1 μm thick and are highly hydrated, the voluminosity of which is strongly dependent upon pH.[14] As the pH is reduced so the internal molecular repulsive forces are reduced, the micelle contracts and the voluminosity declines. In the practical application in question, as the pH of homogenized dairy cream is reduced so the whipping time declines (Fig. 9). One suggested explanation is that pH

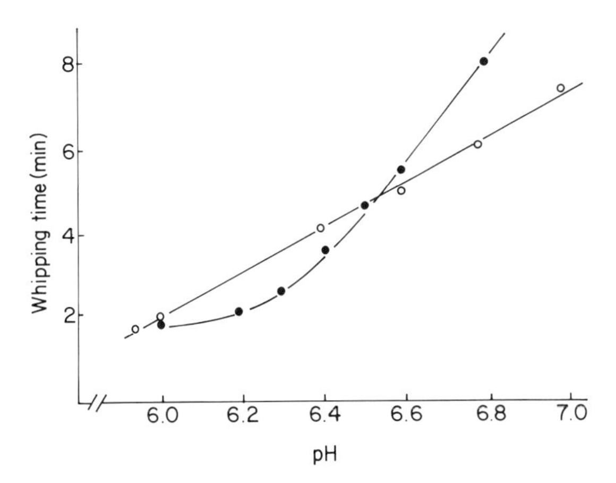

Fig. 9 The effect of pH on the whipping time of homogenized creams. Two different cream sources were used (open and closed circles)

reduction results in a thinner interfacial film between globule and air bubble, hence a higher probability of film rupture and shorter whip times. This is not the complete story because other parameters are also affected by pH which could lead to reduced emulsion stability (e.g. lower protein solubility). Nevertheless, there are strong indications that it is a contributory mechanism.

4 AGGREGATION PROCESSES

Aggregation is a term used to describe a variety of particle-particle interaction processes, the two most important being flocculation and coalescence. There are many texts dealing with both subjects and in this section only a brief summary of the principal features will be given. Flocculation can be represented by a bi-particle process and is therefore frequently found to be second order whilst coalescence (like adsorption) is a uni-particle process and independent of particle concentration.

A Kinetics of flocculation

Where aggregation involves the collision between two particles the rate at which collisions occur and the effectiveness of the collisions will determine the rate of flocculation. The collision frequency is dependent upon particle concentration and the efficiency with which collisions lead to flocculation is related to the energy barrier surrounding the emulsion droplet. The most well known theory of flocculation kinetics is that of Smoluchowski (see ref. 30) which is the basis of many more sophisticated approaches to bi-particle interaction processes.

The simplest description for the flocculation process is given by:

$$\frac{\mathrm{d}N}{\mathrm{d}t} = \alpha . \beta . N^2 \tag{17}$$

where N represents the particle concentration at time t, β is a collision frequency parameter and α is a collision efficiency parameter.

For electrostatically stabilized emulsions it can be shown that to a first approximation α is described by the relationship:

$$\frac{1}{\alpha} = 2 \int_2^\infty \exp(V/KT) \frac{\mathrm{d}z}{z^2} \tag{18}$$

β on the other hand depends upon the mechanism by which encounters occur. Under Brownian motion (perikinetic flocculation)

$$\beta = 4Dd$$

whilst under simple shear flow (orthokinetic flocculation)

$$\beta = \tfrac{2}{3}Gd^3$$

Where G is shear rate, D is particle diffusion coefficient and d is particle diameter. V/KT is the height of the potential energy barrier between two droplets and z is the separation distance between two colliding droplets.

The relative importance of orthokinetic to perikinetic flocculation is proportional to d^2 and is hence strongly dependent on particle size. In practical food emulsions, droplets greater than a few microns flocculate primarily by an orthokinetic mechanism. Even in the absence of an applied shear field, natural convection currents and gravity creaming are sufficient to promote the orthokinetic process.[30]

Whilst there are few practical examples which are absolutely obeyed by this theory it is instructive to appreciate its simplicity. From a measure of the time-dependent change in particle concentration it is possible, in principle, to determine the interaction energy between two colliding particles. Whilst the Smoluchowski approach has been used on many occasions to test the DLVO theory of colloid stability, there are many sources of likely error. Unfortunately, as far as the food scientist is concerned, it is impossible to derive a valid potential energy term for a practical food emulsion.

On the other hand, from an application viewpoint this equation has an important use. If β is constant then a unique measure of emulsion stability is given by α which can be used to compare processing and ingredient effects in a systematic manner with a common reference point. Alternatively, if the parameter α is constant then β can be used to compare different environmental conditions such as shaking, stirring, centrifugation and so on.

The only requirement in terms of application is that

$$\frac{dN}{dt} \alpha N^2 \quad \text{or} \quad \left(\frac{1}{N} \alpha t\right)$$

One of the inherent assumptions in equation 17 is that particles are considered as points in space and that their size does not influence either α or β other than in the manner shown. As particles aggregate so their size changes and this can have significant consequences on both α and β. In practice, therefore, $1/N$ is frequently not proportional to t, even though a second order process is operative.

1 HYDRODYNAMIC CONSIDERATIONS IN FLOCCULATION

When two particles approach one another there are so-called hydrodynamic interactions which diminish the frequency of collision. These hydrodynamic effects relate to the flow behaviour of the continuous phase around the

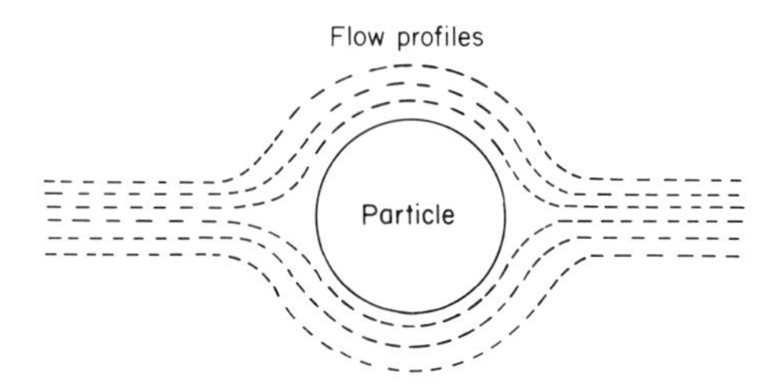

Fig. 10 Schematic illustration of hydrodynamic interaction between two approaching particles. Dotted lines illustrate typical flow profiles for approaching particles with different initial trajectories

surface of the particles. Figure 10 schematically illustrates the flow pattern for one particle interacting with another. The dotted lines represent flow profiles for different particles depending upon the initial trajectory of the approaching particle. This hydrodynamic effect was first considered by Spielman[45] and later by Honig *et al.*[25] in the context of colloid stability. They showed that by including a viscous interaction term previous calculations could overestimate the flocculation rate constant by up to a factor of 10.

Simple viscous interactions between particles of finite size can now be incorporated into theoretical calculations. The simplest form of this interaction is expressed in terms of a reduction in the collision efficiency factor p. Thus Honig *et al.*[25] derived an approximation for the interacting between two equal sized spheres as:

$$\alpha p \approx \alpha\left(\frac{6u^2 + 4u}{6u^2 + 13u + 2}\right) \tag{19}$$

where p is the collision efficiency factor as described by Smoluchowski.

2 ORTHOKINETIC FLOCCULATION

In the majority of practical food systems, the flocculation of emulsion droplets will be governed by an orthokinetic mechanism, either convective, centrifugal or by an external shear force. Under these circumstances it is necessary to be able to describe the collision frequency as a function of the appropriate mechanism. Van de Ven and Mason[53,54] in a series of publications have described the behaviour of a flocculating dispersion of colloidal particles in a shear flow. One of the essential features of their work was to take account of hydrodynamic interactions and particle trajectory effects due to shear flow. In doing so, the collision efficiency factor α also becomes dependent upon G, the shear rate.

Spielman[46] summarizes the work on hydrodynamic effects and in particular reviews the effects of shear on flocculation kinetics. Exact solutions to the flocculation kinetics are only available for the most simple cases of a shear field (laminar flow). Attempts to account for the effect of complex shear fields (e.g. turbulent flow) on flocculation processes have been semiquantitative and can be summarized as follows:

$$\beta = Cd^3\left(\frac{\varepsilon}{\nu}\right)^{1/2}$$

where C is ~ 0.3, ε is the rate of energy dissipation per unit mass and ν is the kinematic viscosity. The effect of turbulence on α is unknown.

For most practical purposes α and β are system related parametric variables which have to be determined empirically and in many cases will be inseparable. It is therefore essential that when using results from a kinetic experiment the conditions under which collisions are being induced are carefully described.

One of the few notable studies on orthokinetic flocculation in a food emulsion was by Van Boekel[4] and Van Boekel and Walstra[5] in which they studied the stability of o/w food emulsions under conditions of laminar and turbulent flow. In particular, they demonstrated that in creams and similar o/w emulsions the presence of crystalline fat had a pronounced effect on the orthokinetic flocculation rate constant which was attributed to the crystals piercing the membrane between two droplets. In the absence of crystalline material the collision efficiency factor was hardly affected by a shear field, whereas in the presence of crystals it was increased by several orders of magnitude.

3 POLYDISPERSITY EFFECTS AND THE APPLICATION OF FLOCCULATION THEORIES

As theoretical descriptions of flocculation processes develop so our appreciation of the complexity of the system also increases. As the theories become more complex, and more precise, so their general applicability to practical emulsion systems diminishes. Whilst it is necessary to appreciate all the limitations of a theory and its application, experimental observations still have to be explained and exploited in a practical manner.

Most emulsions are highly polydisperse in nature, sizes ranging from a few tenths of a micron up to about 1 mm in some cases. All simple theories of flocculation as summarized above, assume a particle size independence of both collision efficiency and frequency. In practice, the particle-particle interaction energy is size dependent (see P. Walstra, this volume), the hydrodynamic interactions are size dependent, particle velocity depends upon mass and hence size. It is not surprising therefore that polydispersity effects can play an important role in flocculation processes.

Polydispersity effects can be accounted for by using a population mass balance approach. In summary, this is simply a formalism to say that the total mass of the system remains the same, only the distribution changes. When applied to flocculation kinetics α and β are required as a function of particle size. The general expression for such a process can be represented by:

$$\frac{\mathrm{d}N_k}{\mathrm{d}t} = \sum_{\substack{i=1 \\ j=k-1}}^{i=k-1} \alpha_{ij}\beta_{ij}N_iN_j - 2\sum_{i=1}^{\infty} \alpha_{ij}\beta_{ij}N_iN_k \tag{20}$$

where i, j and k refer to size classes within a distribution. Thus the rate of change of concentration of particles in size class k is given by the sum of all possible combinations of i and j (where $i + j = k$) minus the sum of k and i combination (i.e. the interaction between k-fold particles and any other particle). This differential approach can then be applied to all size classes within the distribution to account for the total population of particles.

If α_{ij} and β_{ij} are known for all possible configurations it is possible to solve the complete set of differential equations numerically. Unfortunately it is not easy to do the reverse, determine α_{ij} and β_{ij} from an observation of the change in particle size distribution as a function of time.

This approach has been used, for example by Suzuki *et al.*[49] to predict the particle size distribution changes in emulsions undergoing perikinetic flocculation using calculated values for α and β. From the computer simulation experiments the total particle concentration

$$\left(\sum_{k=1}^{\infty} N_K\right)$$

can be calculated and compared with the simple Smoluchowski predictions. Figure 11a and b illustrate how, by taking into account the effects of polydispersity, the linearity of $1/N$ vs t is rapidly lost.

In these examples Suzuki *et al.*[49] chose three possible scenarios for α (the collision efficiency factor) 1, $\leqslant 1$, and $\ll 1$ (so called rapid, rapid/slow and slow flocculation conditions). Figure 11a refers to an initial situation consisting of monodisperse particles and under these circumstances only slow flocculation deviates significantly from the simple Smoluchowski model. Figure 11b refers to an initially polydisperse system and in this case both rapid/slow and slow flocculation processes are significantly affected by polydispersity.

The implications of these model calculations which take polydispersity into account is that for a relatively stable emulsion (which includes most food emulsions) the simple second order kinetic plot should not be linear.

There are many additional complications other than polydispersity. Most

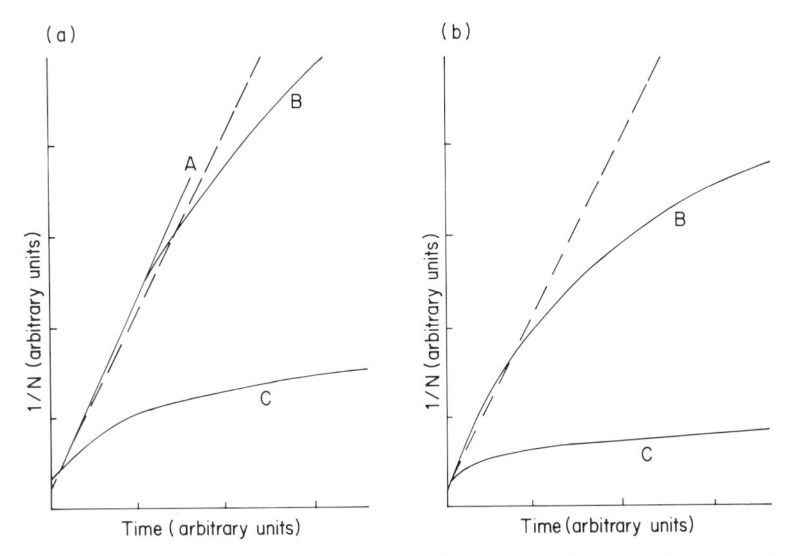

Fig. 11 The effects of polydispersity on the time course of flocculation as predicted by Smoluchowski kinetics. Dotted line assumes no polydispersity effects. A, rapid. B, rapid/slow and C slow flocculation. (a) For an initially monodisperse system. (b) For an initially polydisperse system. (Reproduced with kind permission from [49])

theories, for example, assume simple bi-particle collisions to occur. In practice, of course, food emulsions frequently contain high disperse phase volumes and multibody collisions can occur.

Perhaps one of the most serious problems with all kinetic theories is the calculation of an interaction energy term between two colliding droplets. This can really only be done for simple, well defined, colloids.

If particles flocculate, but do not coalesce, then the flocculated species will not be spherical and all previous assumptions pertaining to spherical particles may be invalidated. Furthermore the flow behaviour of an irregular particle, and hence hydrodynamic effects, will be difficult to quantify.

B Coalescence process

Coalescence refers to the rupturing of the interfacial film between two adjacent droplets and the loss of integrity of the separate individual droplet identities. It has many similarities to the adsorption process and can be described by a first order kinetic equation of the type:

$$N_t = N_0 e^{-Kt} \tag{21}$$

where K is related to the probability of the interfacial film rupturing in time t.

The process is stochastic and can be treated using probability theory. Unfortunately all attempts to relate K to some quantitative combination of film properties has failed.

It has been considered that interfacial rheology is an important factor influencing the rupturing process. It is true to say that correlations do exist between coalescence stability and interfacial rheology[2,3,6] but here are many more examples where no correlation exists at all. Graham and Phillips[21] concluded from their work on proteins at interfaces, that rheological properties were not predominant in determining coalescence. They considered that film thickness and disjoining pressure may be the critical factors. In the cases where correlations exist between interfacial rheology and stability one can also argue that film thickness is changing.

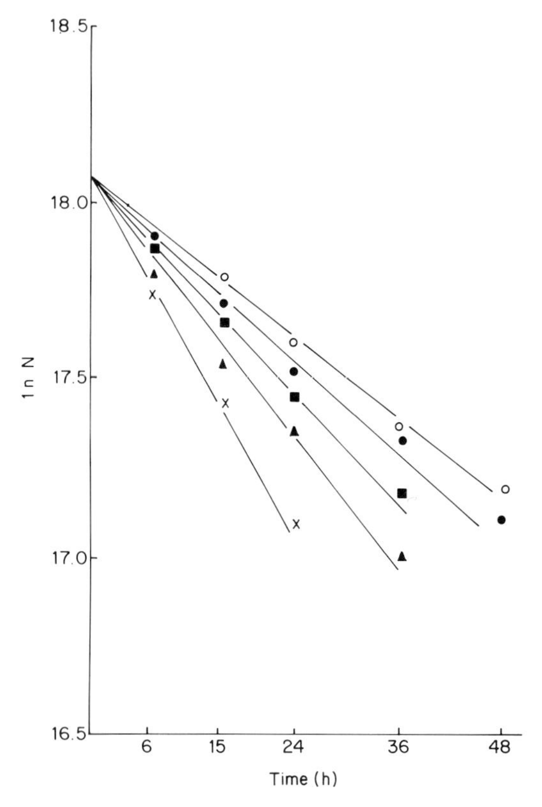

Fig. 12 The effect of phenol on the coalescence rate of bovine serum albumin stabilized o/w emulsions. Phenol concentrations: ○ zero. ● 0.005 M. ■ 0.01 M. ▲ 0.02 M. × 0.04 M. (Reproduced with kind permission from [47])

To the author's knowledge the variables have hitherto not been separated, and therefore no definitive conclusions can be drawn.

Srivastava[47] reported that coalescence rate was related logarithmically to surface elasticity. These observations are based upon the use of small molecule surfactants to decrease the surface elasticity of protein stabilized emulsions.

Figure 12 illustrates how the addition of phenol increases coalescence rate for a BSA stabilized emulsion. Other workers, for example, Pearson[31] and Cumper and Alexander[12] have observed similar trends for the effect of small molecule surfactants. Pearson, for example, demonstrated that 0.05 M decyltrimethylammonium bromide addition to a β-lactoglobulin or BSA stabilized emulsion increased coalescence rates by about a factor of 100. Cumper and Alexander also observed that oleyl alcohol substantially decreased the coalescence stability of protein stabilized emulsions.

An alternative explanation (to surface rheology) for these small molecule surfactant effects is their ability to displace proteins at interfaces because they exert a higher surface pressure (Darling and Birkett, to be published).

If proteins are displaced the film thickness is diminished and hence the equilibrium distance of closest approach between flocculated droplets, which precedes coalescence, is also reduced. In nearly all cases where macromolecular stabilizers are used, one can explain qualitatively the reduction in stability in terms of some effective film thickness terms (as yet unquantified). This reduction is often brought about as a result of protein displacement by small molecule surfactants exerting a higher surface pressure.

To the author's knowledge the parameters governing film rupture have never been independently varied and hence definitive conclusions cannot be drawn. There are a variety of phenomenological descriptions of the coalescence processes, but there is no generally accepted theory that relates the rate constant (K in equation 21) to the fundamental properties of the interface.

C Application of simple kinetic theories

Simple kinetic theories should not apply to practical emulsions, as reasoned in the foregoing discussions. However, despite the various uncertainties, simple kinetic models can be successful provided the results are not over-interpreted.

Sherman[40] for example has described the kinetics of coagulation of model w/o and o/w food emulsions using a two stage first order kinetic model. Figure 13 illustrates the linear relationships between log (particle concentration) and time. Whilst the results are adequately described by his first order kinetic model the subsequent interpretation should be treated cautiously. In this particular example, the two stages are reported to relate to

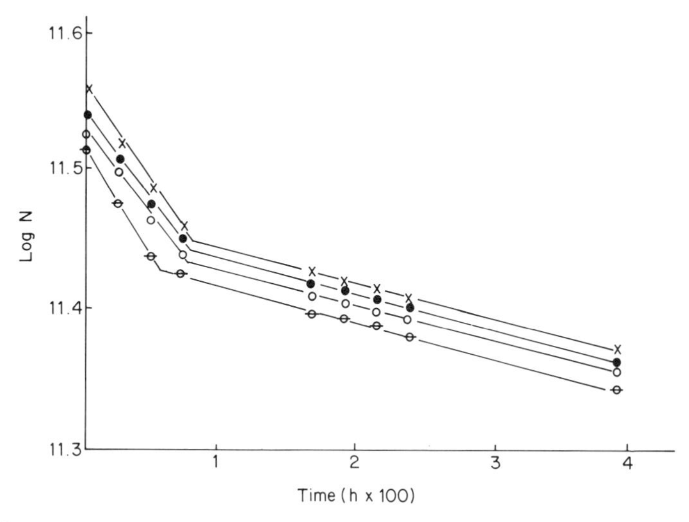

Fig. 13 Coalescence of water in paraffin oil emulsions (63% dispersed phase) stabilized by sorbitan mono-oleate. First order kinetic plot. Emulsifier concentration. × 6.0%. ● 4.5%. ○ 3.0%. ⊖ 1.5%. (Reproduced with kind permission from [40])

the different stabilities of small and large particles. Whilst this may be a contributory factor, it is essential to consider all possible factors likely to influence the kinetic process before attempting a mechanistic interpretation. With w/o emulsions for example, the finite solubility of water in oil can lead to disproportionation (Ostwald ripening) which would lead to a preferential loss of small droplets. Thus, within experimental error, the initial stage of the data in Fig. 13 could be due to disproportionation and, in fact, only one aggregation process may be operating.

Careful use of control experiments can often eliminate alternative mechanisms. In the above case, for example, addition of salt to the water phase of a w/o emulsion would inhibit Ostwald ripening by providing a counteracting osmotic effect. Thus the contribution of a disproportionation process can be eliminated.

In other experiments, the data may be too imprecise for a definitive kinetic model to be established. For example, in studies by the author on emulsions stabilized with Tween 60/Span 60 (Fig. 14) the particle concentration data were found to be consistent with a second order kinetic process, but the error in measurement is too large to be confident that a bi-particle aggregation mechanism is operative. Nevertheless, in these practical examples the apparent rate constant can still be used to summarize the data and illustrate important differences in emulsion behaviour.

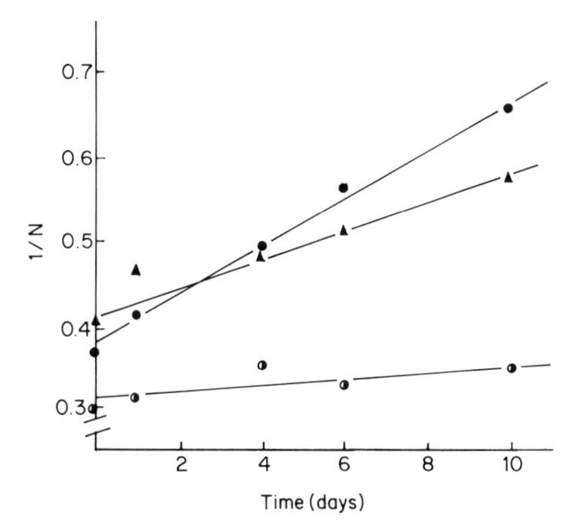

Fig. 14 Second order kinetic plot for the change in particle concentration in a model food emulsion as a function of the method of emulsifier addition. Emulsifiers used 0.2% Tween 60 and 0.2% Span 60. ● both emulsifiers added to the oil phase. ▲ both emulsifiers added to the aqueous phase. ◑ Tween 60 in aqueous phase, Span 60 in oil phase. Oil phase 20% Palm Kernel Oil (slip point 39°)

The data in Fig. 14, for example, illustrate how the method of incorporating the emulsifier can influence stability (an observation which, in process terms, can have significant implications for product properties). The kinetic models are convenient descriptors of emulsion behaviour provided the causality of the observations is not inferred from an unsound mechanistic interpretation of the data.

In studies on the colloidal stability of ice cream premix emulsions, Sherman[40] used this phenomenological approach to show that milk proteins were the main contributor to emulsion stability and that the small molecule emulsifier (monoglyceride) had a synergistic effect towards stability. The mechanism by which this occurs cannot, however, be inferred from the results.

The whipping of cream is another example of a kinetic phenomenon. It is a complex set of events involving flocculation, coalescence and adsorption processes. Nevertheless, Shioya *et al.*[41] have successfully described cream whipping using a simple first order kinetic model according to the equation:

$$\ln \eta_t/\eta_0 = K_1 t + C_1 \tag{22}$$

$$\ln(\eta_t/\eta_0 - \exp(K_1 t)) = K_2 t + C_2 \tag{23}$$

where η_0 and η_t are the initial viscosity and viscosity after time t respectively;

K_1 and K_2 are first order rate constants. Figure 15a and b illustrate the application of this model to the viscosity changes occurring during cream whipping. A good correlation between K_1 and K_2 with, respectively, liquid cream stability and cream whippability was observed (Fig. 16a and b). Such correlations enable the physical properties of the creams to be predicted from simple kinetic measurements and in this context provide a valuable practical tool to the product technologist.

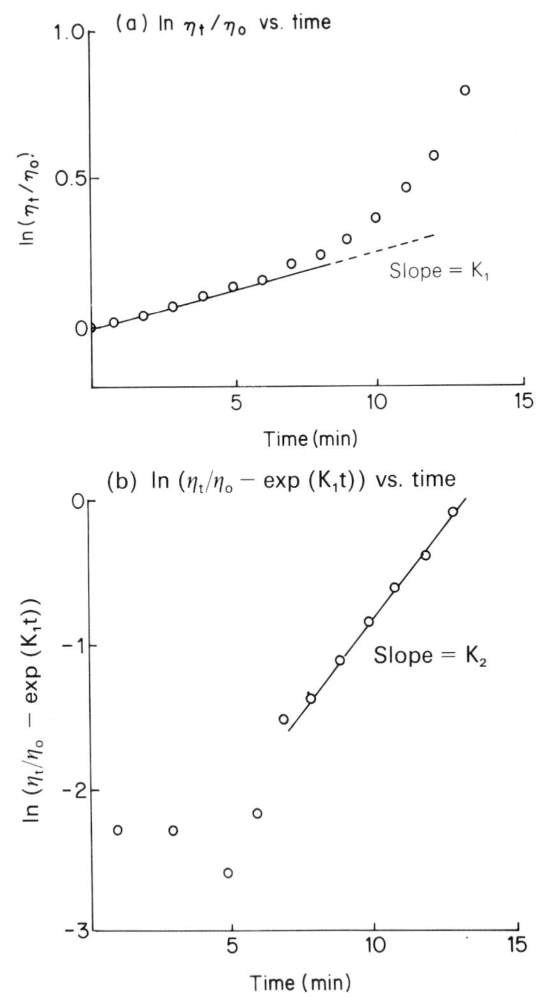

Fig. 15 Effect of shear on viscosity of dairy creams as a function of time. (Reproduced with kind permission from [41])

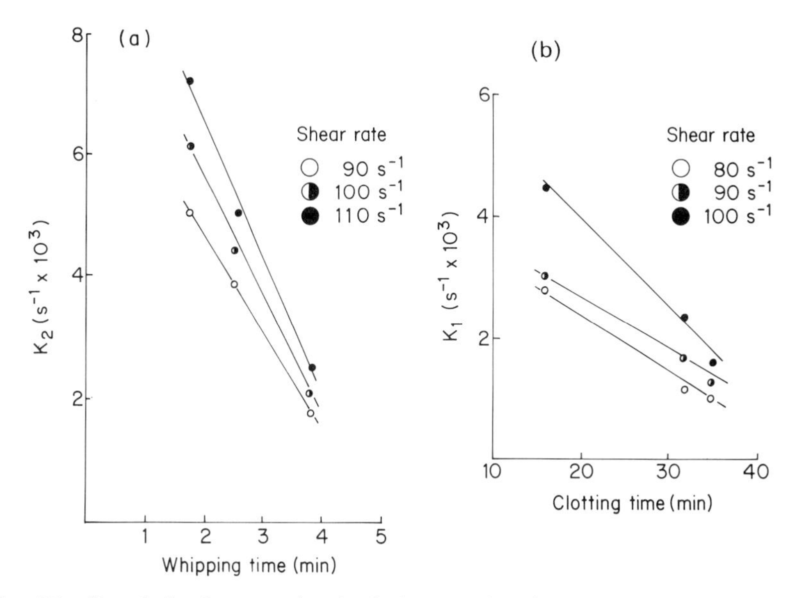

Fig. 16 Correlation between the physical properties of dairy cream and aggregation rate constants. (a) Whipping time and K_2 (see text). (b) Clotting time and K_1 (see text). (Reproduced with kind permission from [41])

The author and coworkers have used a kinetic model of aggregation phenomena to describe the physical stability of emulsions in a more general manner.

This model describes three kinetic processes—flocculation, deflocculation and coalescence according to the reaction scheme.

$$\bigcirc + \bigcirc \underset{K_{-1}}{\overset{K_1}{\rightleftharpoons}} \bigcirc\!\!\bigcirc \overset{K_2}{\rightarrow} \bigcirc\!\!\bigcirc$$

$[1-x]$ $[x-y]$ $[y]$

No attempt is made at defining the physical state of either flocculated or aggregated species. Suffice it to say that the flocculation process is reversible, whilst coalescence is not. The model includes a variety of inherent assumptions, as typified by some of the points raised in Section 4.A.3, not least the fact that all particles involved in either flocculation or coalescence do so independently of particle size.

Fundamentally, the model does not describe the absolute physical phenomena. However, the three rate constants can be equated with practical observations on emulstion stability and in turn related to the effect of process and composition. Thus they can have both conceptual significance and practical application.

1 APPLICATION OF THE GENERALIZED KINETIC MODEL

If at any instant in time the concentration of primary, flocculated, and coalesced species is given by $[1-x]$, $[x-y]$ and $[y]$ respectively, the differential equations describing the rate of change of concentrations according to the above kinetic model are given by:

$$\frac{d[x]}{dt} = K_1[1-x]^2 - K_{-1}[x-y] \tag{24a}$$

$$\frac{d[y]}{dt} = K_2[x-y] \tag{24b}$$

The solution to the above differential equations requires numerical integration and is impractical for product application purposes. Furthermore, since the model has little fundamental significance, absolute solutions are unnecessary. A practical solution can be obtained by truncating the quadratic term in equation (24a) which, provided $x \lesssim 0.1$, can be shown to be a valid approximation.

Equation (24) then leads to:

$$\frac{d[x]}{dt} \simeq K_1[1-2x] - K_{-1}[x-y] \tag{25}$$

Through the use of standard texts on differential equations the solution to equations 24 and 25 can be shown to be:

$$[x] = A\underbrace{(1-\exp(-\lambda_1 t))}_{(P)} + B\underbrace{(1-\exp(-\lambda_2 t))}_{(Q)}$$

$$[y] = \frac{2K_1 + K_{-1} - \lambda_1}{K_{-1}}(P) + \frac{\lambda_2 - (2K_1 + K_{-1})}{K_{-1}}(Q)$$

where λ_1 and λ_2 are the negative and positive roots respectively to the quadratic equation:

$$\lambda^2 - (2K_1 + K_{-1} + K_2)\lambda + 2K_1K_2 = 0$$

$$A = \frac{1}{2}\left(\frac{\lambda_2 - 2K_1}{\lambda_2 - \lambda_1}\right)$$

$$B = \frac{1}{2}\left(\frac{2K_1 - \lambda_1}{\lambda_2 - \lambda_1}\right)$$

In practice, one or perhaps two rate constants are frequently insignificant and further simplifications can be made. For example, when $k_1 \gg k_{-1}$ or k_2 the model approximates to second order, irreversible flocculation with no coalescence, whereby:

$$\frac{1}{[1-x]} = K_1 t + \text{constant} \tag{26}$$

If $k_2 \ll k_1$ or k_{-1}, the model reduces to reversible flocculation which has a solution for x in terms of an equilibrium concentration given by:

$$\ln \frac{x_e(1 - x x_e)}{x_e - x} = \frac{K_1(1 - x_e)^2 t}{x_e} \tag{27}$$

Finally if $k_1 > k_2 > k_{-1}$ then the solution approximates to first order kinetics and the solution is given by

$$\ln[x] = K_1 t + \text{constant} \tag{28}$$

Only when all three rate constants are of the same order of magnitude is it necessary to use the complete solution of equations 24 and 25 and for the majority of practical purposes one of the simple approximate solutions will suffice.

In many o/w food emulsions, where the fat phase is partially crystalline, the emulsion droplets retain their identity when aggregation occurs, even when the intervening film between droplets is ruptured. Under these circumstances a practical method of estimating the degree of aggregation is by viscometry.

2 RHEOLOGICAL STUDIES ON KINETICS OF AGGREGATION

In a study of the reversible flocculation of emulsion droplets Van den Tempel[51] used steady-state viscosity measurements to estimate the degree of flocculation. Using an empirical relationship between viscosity and phase volume, as described by Mooney,[28] the effective dispersed volume in a flocculated emulsion is given by:

$$\theta = vN_1 + v.f.(N_0 - N_1)$$

where v is the primary particle volume, N_0 is the total number of primary particles and N_1 the concentration of primary particles left unaggregated under steady state conditions; f is a swelling factor which accounts for the occluded liquid between particles within an aggregate (typically f is around 1.6).

By the use of an extended Smoluchowski approach for flocculation kinetics and taking account of different size aggregates, van den Tempel was able to equate the steady-state viscosity of an emulsion to a particle-particle interaction parameter and thus obtain a quantitative measure of stability.

De Vries[57] has used a simpler approach to study the shear induced aggregation of polymer lattices under a variety of conditions. He observed an apparent aggregation time where the viscosity suddenly increased markedly. He used the reciprocal of this aggregation time as a measure of aggregation rate making the assumption that a single-rate constant could describe the total process.

In a more recent series of papers Hattori and Izumi[22] describe the relationship between coagulation rate theory and viscosity using DLVO theory as the descriptor of aggregation. Their approach included both perikinetic and orthokinetic flocculation as well as deflocculation induced by shear. As a result, they were able to describe, at least qualitatively, complex hysteresis effects in the rheology of aggregating suspensions.

Finally, with the aid of the kinetic model described by equations 24 and 25, the author has used a rheological technique to describe the shear induced aggregation of cream emulsion droplets. Using a similar principle to Van den Tempel,[51] Darling *et al.* (to be published) have determined the degree of aggregation from an apparent phase volume calculated from measurements on viscosity. The kinetic rate constants were obtained from computer curve fitting between viscosity-time output and the predicted rate equation.

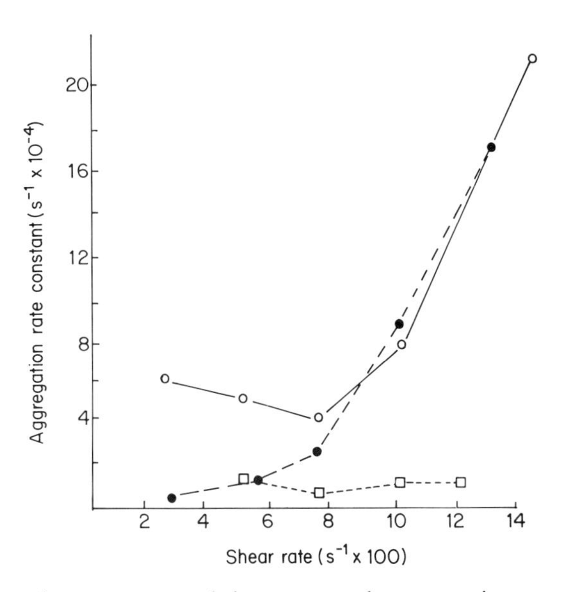

Fig. 17 Effect of temperature and shear rate on the aggregation rate constant for a model food. ○ 20 °C; ● 15 °C; □ 10 °C. Emulsion formulation: 40% Palm Kernel Oil (slip point 39°C) in 0.6% solution of sodium caseinate. Particle size. $d_{32} = 1.04$ μm

As an example of the application of this approach Fig. 17 illustrates the effect of shear rate and temperature on the stability of a model o/w food emulsion. In this particular example, deflocculation was negligible and the kinetics approximated to a second order process. It is clear from Fig. 17 that temperature has a pronounced effect on the emulsion stability. This is consistent with the churning properties of cream where an increase in temperature (up to ~30 °C) leads to a rapid reduction in churning time. The rate controlling step in this case is coalescence (k_2) and as the solid fat content diminishes, so the rate at which droplets coalesce increases.

The main advantage of the viscometry technique in determining kinetic rate constants is in avoiding the need to measure directly either particle concentrations or particle size, both of which are a significant source of experimental error and can often lead to erroneous conclusions about mechanisms.

5 SOME THEORETICAL CONSIDERATIONS OF COMPLEX KINETIC EVENTS

In the practical world of food emulsions where kinetic events are always present, simple kinetic theories, for the reasons given earlier, rarely describe experimental observations accurately. There have been a few attempts to study complex kinetic phenomena in dispersions where two or more processes are operating either consecutively or simultaneously. These will now be briefly discussed.

So far, four separate processes have been considered (separation, adsorption, flocculation and coalescence). There are therefore 6 pairwise, 4 three-way, and one four-way interactions possible.

A Aggregation and separation

There are many instances where an emulsion undergoes flocculation or coalescence and the aggregated particles subsequently cream due to their increase in size. Under these circumstances the rate of creaming is enhanced over that observed with no aggregation. Since droplets are creaming, both perikinetic and orthokinetic events have to be considered.

Reddy *et al.*[33] have calculated, using the population mass balance approach, the change in particle size distribution at any vertical position in an emulsion undergoing flocculation and creaming.

The flocculation mechanism was based upon DLVO theory and creaming was considered to be described by Stokes' equation. Having set up the population balance the predictions were compared with experimental observations[32] on a model paraffin /water emulsion stabilized with an ionic surfactant. An example of these results is given in Fig. 18a and b. Considering the complexity of the problem this represents an extremely good

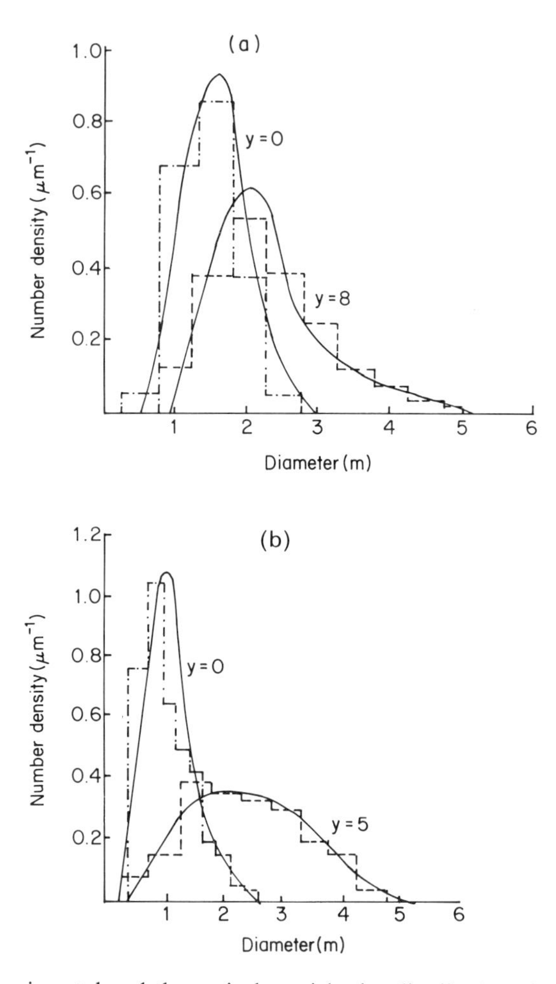

Fig. 18 Experimental and theoretical particle size distributions for paraffin/water emulsions undergoing aggregation and separation. (a) Particle size distributions after 3 h at base of container ($y = 0$) and 8 cm above base ($y = 8$). Histograms, experiment. Full curves, theory. (b) Particle size distributions after 4 h at base of container ($y = 0$) and 5 cm above base ($y = 5$) (see ref. for experimental details). (Reproduced with kind permission from [32])

fit between theory and experiment. These data show the particle size distribution as measured by electron microscopy compared with the computer predictions for two segments of the container.

B Adsorption and aggregation

Particle adsorption processes are generally associated with aggregation in food emulsions; notable examples being the whipping and churning behaviour of cream and the aeration of ice cream. Whilst there has been no fundamental work on emulsions involving adsorption and aggregation, Heller and Peters[23] developed a semiquantitative approach to surface coagulation of $Fe(OH)_2$ particles. They showed that surface coagulation proceeded faster than the equivalent bulk process and that the coagulation rate could be described by the following equation:

$$\frac{-\mathrm{d}[x]}{\mathrm{d}t} = k_0\left(\frac{s}{v}\right)\frac{[x]^2}{[a + bx]^2}$$

where x is the bulk colloid concentration, k_0 is the rate constant and a and b are constants from a Langmuir isotherm approach. $\frac{s}{v}$ is related to the surface to volume ratio of the gas bubbles.

The approach assumes that the rate controlling step is coagulation and that adsorption is sufficiently rapid not to influence the coagulation rate. This technique has not been applied although it would be directly relevant, for example, to cream whipping.

6 CONCLUDING REMARKS

In the foregoing text a summary of the kinetic behaviour of colloidal dispersions has been given paying particular attention to the relationships between rate theories and experimental observations. Unfortunately, there are few practical examples in the area of food emulsions where theory and experiment can be equated quantitatively. The problem lies in the fact that food emulsions are so complex and no theory is capable of predicting absolute results in any but the simplest of cases.

Foods, which are rarely equilibrium structures, are continually changing in so many ways that it is foolhardy to expect a simple quantitative theory to explain the kinetics of change. A more immediate and very important objective is the ability to model kinetic processes such that self consistent changes can be predicted on the basis of a set of limited observations. This is the area of kinetic modelling. In the author's opinion the purpose of a kinetic model is to describe in a concise and conceptually relevant form, all experimental observations within a data set. In addition a kinetic model should, wherever possible, aid interpretive reasoning about the behaviour of the system. It should not in itself be a means of determining the mechanism underlying the kinetic change. On the basis of the above argument it is not essential for a model to be absolute. Provided it is self consistent, and can predict within its limiting boundary conditions it has its uses.

In applying a kinetic model it is essential that the product technologist appreciates the limitations of the model; only then can it be constructively used to aid technological progress. Unfortunately kinetic models are frequently used without appreciating the underlying assumptions on which they are based.

It has been the intention within this chapter to outline the kinetic theories available to the food emulsion scientist and illustrate their application wherever possible. By describing instability of emulsions in terms of kinetics rather than thermodynamic or single point measurements, the author believes that a more unique and meaningful appreciation of food emulsions will be achieved.

Glossary

a	Acceleration.
α	Collision efficiency between particles undergoing flocculation.
αp	Collision efficiency factor incorporating hydrodynamic interactions.
β	Collision frequency between particles.
C	A constant.
d	Particle diameter.
D	Particle diffusion coefficient.
e	Voidage ($1-\theta$, where θ represents disperse phase volume).
e_l	The value of e when $V_n(1 - x_1)$ is a maximum.
ε	The rate of energy dissipation per unit mass.
f	Swelling factor for particles in an aggregated suspension.
g	Acceleration due to gravity.
G	Shear rate.
γ	Interfacial energy (subscripts refer to appropriate phases being considered).
h	Film thickness.
H	Average thickness of a drop spreading at an interface.
i, j, k	Subscripts—refer to specific size classes in a polydisperse system.
$J(p)$	Trigonometric expression for penetration depth of a droplet into an interface.
k	A rate constant (subscripts refer to specific processes).

n An exponent.
N Number concentration of droplets.
N_0 Initial number concentration of droplets.
ν Kinematic viscosity.
r Radius.
ρ Density.
$\Delta\rho$ Difference in density between two phases.
s Surface area of dispersed phase.
S Spreading coefficient for droplet at an interface.
t_{12} Time required for a film to thin between two points.
t Time.
T Temperature,
u Relative velocity between two particles undergoing hydrodynamic interaction.
V_h Hindered settling particle velocity.
V_s Calculated particle velocity according to Stokes' equation.
V Maximum repulsive potential energy term.
v Volume of dispersed phase.
w Angular velocity.
W Work of adhesion between droplet and interface.
x Fraction of particles involved in aggregation (flocculated and coalesced).
y Fraction of particles involved in coalescence.
z Distance between two approaching particles.

References

1. Becher, P. (Ed.) (1983). Encyclopaedia of Emulsion Technology.
2. Biswas, B. and Haydon, D. A. (1962). *Kolloid Z.* **185**, 31.
3. Biswas, B. and Haydon, D. A. (1962). *Kolloid Z.* **186**, 57.
4. van Boekel, M. A. J. S. (1980). PhD Thesis, Agricultural Research Reports, Wageningen, ISBN, 90-220-0739-1.
5. van Boekel, 7. Brinkman, H. C. (1947). *Applied Sci. Res.* **A1**, 27.
6. Boyd, J., Parkinson, C. and Sherman, P. (1972). *J. Colloid Interface Sci.* **41**, 359.
7. Brinkman, H. C. (1947). Applied Sci. Res. **A1**, 27.
8. Brinkman, H. C. (1948). *Applied Sci. Res.* **A1**, 81.
9. Brinkman, H. C. (1949). *Applied Sci. Res.* **A2**, 190.
10. Charles, G. E. and Mason, S. G. (1960). *J. Colloid. Sci.* **15**, 236.
11. Cockbain, E. G. and Roberts, T. S. (1953). *J. Colloid. Sci.* **8**, 440.
12. Cumper, C. W. N. and Alexander, A. E. (1950). *Trans. Farad. Soc.* **46**, 235.
13. Darling, D. (1982). *J. Dairy Res.* **49**, 695.
14. Darling, D. F. (1982). The Effect of Polymers on Dispersion Properties, ed. Th. Tadros, Academic Press, London.
15. Darling, D. F. and Butcher, D. W. (1978). *J. Dairy Res.* **45**, 197.
16. Davies, L., Dollimore, D. and Sharp, J. H. (1976). *Powder Tech.* **13**, 123.

17. Davies, L., Dollimore, D and McBride, G. B. (1977). *Powder Tech.* **16**, 45.
18. Dickenson, E. (1980). *J. Colloid Interface Sci.* **73**, 578.
19. Fox, K. K., Holsinger, V. H., Caha, J. and Pollansch, M. J. (1960). *J. Dairy Sci.* **43**, 1396.
20. Gillespie, T. and Rideal, E. K. (1956). *Trans. Farad. Soc.* **52**, 173.
21. Graham, D. E. and Phillips, M. C. (1976). Theory and Practice of Emulsion Technology, ed. A. L. Smith, Academic Press, London.
22. Hattori, K. and Izumi, K. (1982). *J. Dispersion Sci. Tech.* (a) **3**, 129; (b) 147; (c) 169.
23. Heller, W. and Peters, J. (1970). *J. Colloid Interface Sci.* (a) **32**, 592; (b) **33**, 578.
24. Higuchi, T. (1958). *J. Am. Pharm. Ass.* **47**, 657.
25. Honig, E. P., Roeberson, G. J. and Wiersema, P. H. (1971). *J. Colloid Interface Sci.* **36**, 97.
26. Lang, S. B. and Wilke, C. R. (1971). *Ind. Eng. Chem. Fund.* **10**, 329.
27. Mar, A. and Mason, S. G. (1967). *Kolloid-Z. Z. fur Polymer* **224**, 161.
28. Mooney, M. (1946). *J. Colloid Sci.* **1**, 195.
29. Mulder, H. and Walstra, P. (1974). The Milk Fat Globule, Commonwealth Agricultural Bureaux, Farnham Royal, Bucks.
30. Overbeek, J. Th. G. (1952). Colloid Science, ed. H. R. Kruyt, Vol. 1, p. 278.
31. Pearson, J. T. (1968). *J. Colloid Interface Sci.* **27**, 64.
32. Reddy, S. R. and Fogler, H. S. (1981). *J. Colloid Interface Sci.* **82**, 128.
33. Reddy, S. R., Melik, D. H. and Fogler, H. S. (1981) *J. Colloid Interface Sci.* **82**, 116.
34. Richardson, J. F. and Zaki, W. N. (1954). *Chem. Eng. Sci.* **3**, 65.
35. Shepherd, I. S. and Yoell, R. W. (1976). Food Emulsions, ed. S. Friberg. Academic Press, New York.
36. Sherman, P. (1967). *Proc. 4th Int. Cong. on Surface Active Substances*, Vol. 2, p. 1199.
37. Sherman (1968). *Emulsion Science*, ed. P. Sherman, Academic Press, London.
38. Sherman, P. (ed.) (1968). Emulsion Science. Academic Press, London.
39. Sherman, P. (1969). *J. Texture Studies* **1**, 43.
40. Sherman, P. (1971). *Soc. Pharm. Chem.* **Nov**., 693.
41. Shioya, T., Kako, M., Taneya, S. and Sone, T. (1981). *J. Texture Studies* **12**, 185.
42. Silber, D. and Mizrahi, S. (1975). *J. Food Sci.* **40**, 1174.
43. Smith, L. M., Carter, M. B., Dairiki, T., Acuma-Bonilla, A. and Williams, W. A. (1977). *J. Agric. Food Chem.* **25**, 647.
44. Smith, L. M. and Dairiki, T. (1975). *J. Dairy Sci.* **58**, 1254.
45. Spielman, L. A. (1970). *J. Colloid Interface Sci.* **33**, 562.
46. Spielman, L. A. (1978). The Scientific Basis of Flocculation, ed. K. J. Ives, Sijthoff & Noordhoff. Alphen aan den Rijn.
47. Srivastava, S. N. (1964). *J. Indian Chem. Soc.* **41**, 279.
48. Steinmour, H. H. (1944). *Ind. Eng. Chem.* **36**, 618.
49. Suzuki, A., Ho, N. F. H. and Higuchi, W. I. (1969). *J. Colloid Interface Sci.* **29**, 552.
50. Svarovsky, L. (1982). Encyclopedia of Chemical Engineering Technology, eds R. E. Kirk and D. F. Othmer, Wiley.
51. van den Tempel (1963). Rheology of Emulsions, ed. P. Sherman, Pergamon Press, London.
52. Troy, H. C. and Sharp, P. S. (1929). *J. Dairy Sci.* **195**.
53. Van de Ven, T. G. M. and Mason, S. G. (1977). *Colloid Polymer Sci.* **255**, 468.
54. Van de Ven, T. G. M. and Mason, S. G. (1976). *J. Colloid Interface Sci.* (a) **57**, 505; (b) 517; (c) 535.

55. Vold, R. D. and Groot, R. C. (1962). *J. Phys. Chem.* **66**, 1969.
56. Vold, R. D. and Groot, R. C. (1967). *Proc. of 4th Int. Cong. on Surface Active Substances*, Vol. **2**, p. 1233.
57. de Vries, A. J. (1963). Rheology of Emulsions, ed. P. Sherman, Pergamon Press, London.
58. Walstra, P. and Oortwijn, H. (1975). *Neth. Milk and Dairy J.* **29**, 263.

8 The Basic Principles of Mechanical Failure in Biological Systems

A. G. ATKINS

Department of Engineering, University of Reading

1 INTRODUCTION

This paper* is concerned with the mechanical behaviour of biological materials, e.g. strength, stiffness, crack resistance etc., of living animals and plants, and lessons which may be learnt from study of the "design" of biological materials.[2,3]

2 THE CHEMICAL STRUCTURE OF BIOLOGICAL MATERIALS

Vincent[18] has described animal and plant organisms as "chemical factories" in which synthetic abilities must be optimized, in other words, the successful organism will produce the most effective materials for the least effort, using as few chemical reactions as possible, all in an aqueous environment. It is then possible to rationalize the fact that nature usually employs only two types of

* This chapter is based, in part, on two articles entitled "Biomaterials in Industry" (presented at the 1983 Annual Meeting of the British Association for the Advancement of Science) and "Industrial Lessons from Nature" (published in "Spectrum" by The Central Office of Information).

FOOD STRUCTURE AND BEHAVIOUR
ISBN 0-12-104230-8

polymer out of all those available. They are the proteins (polymers of amino acids) and the polysaccharides (polymers of sugars). Both classes of polymer are extremely adaptable and nature uses them as fibres, space-fillers, binders of other chemicals, rubbers, etc. to produce both compliant and stiff skeletal tissues. Materials ranging from infinitely soluble lubricants (e.g. mucus) to exceedingly stable rubbers (e.g. elastin which can withstand temperatures of 120 °C without damage) are made in this way. In addition, there is often a mineral phase present particularly when compressive strength is required. The mineral phase consists of, for example, calcium salts in the skeletons of animals that live in the sea. Finally, and most importantly, water is present in biological materials; its profound effect on mechanical behaviour makes its presence or absence all the more significant.

Owing to the geometry of the peptide bond there is a limited number of shapes available to proteins. These are mainly helices (one or more chains arranged in an interlocking spiral) and sheets (parallel chains arranged in a planar fashion). Fibres tend to be made of one type of conformation, e.g. keratin fibre (α-helix), collagen (triple helix), silks (antiparallel β-sheets). Sugar units can be bonded together in a greater variety of ways, although their chemical variation is less. They can also form branching polymers since each carbon atom of a sugar (mostly hexoses) can form a bond with another sugar. Sugars can form fibres (e.g. cellulose and chitin) which can be more stable than protein fibres. They can also entrain vast amounts of water (a firm gel can be made with only 2.5% of the right polysaccharide) to form materials capable of taking compressive forces such as those experienced by seaweeds, intervertebral discs and cartilage.

Organic materials are metabolically expensive to produce, and Currey[6] remarks that it is not surprising that many animals (particularly those that live in the sea) have used minerals to produce the stiffness, and in some cases other mechanical properties, often required in skeletons. There are only three widely-used minerals: calcium carbonate, silica and calcium phosphate. These are not minerals that would immediately spring to mind as being the best for the purpose of producing stiff, fairly tough structures. But, we must remember the constraints on animals: all processes must take place within the temperature range 0°–40°, and for most animals the range is between 10°–20°. Also, animals can use only those materials that are immediately available in the environment. Most animals live in the sea, and the most widely available cation is sodium, but this produces very soluble salts. Currey[6] points out that the use of calcium, which produces fairly insoluble salts, but which is nevertheless fairly common in the sea, seems an effective compromise between solubility and availability.

By far the commonest skeletal material is calcium carbonate; it forms the major skeletal mineral of corals, barnacles, molluscs, many sponges etc.

Mollusc shells (snails, mussels, scallops etc.) have calcium carbonate in the crystalline form of either calcite or argonite with, importantly, always a more-or-less tenuous matrix of organic material—mainly protein and a small amount of polysaccharide.

Although the chemical ingredients in biological materials are small in number, the variety of possible combinations is immense. Depending upon the biological requirement, almost any arrangement of components may be found. In studying the make-up of biological materials, it is difficult in many instances to identify "a fibre" or "a matrix" or even "a material" as opposed to "a structure". This traditional kind of classification used by engineers and materials scientists is much more difficult to apply to biology. In most cases one finds a complex hierarchy of structural elements, from single macromolecules to macroscopic units, each level being a composite in its own right. This is true not only in soft, flexible materials such as skin, blood vessels, tendons etc., but also in rigid materials like wood, bone and teeth which conform more to our concept of engineering composites, Fig. 1.

3 REPRESENTATIVE MECHANICAL PROPERTIES

The mechanical properties of biological materials may be determined in the traditional ways employed by materials scientists for metals, polymers and ceramics to arrive at tensile and compressive strengths, elastic moduli, works of fracture and so on. Some biological materials, such as bone, teeth, horn, antler, wood and animal ceramics are more-or-less Hookean elastic and stiff and hard; in this way they behave similarly to most engineering solids, and established methods of analysis may be applied to them. Other biological materials, such as skin, artery, gut and bladder are highly extensible soft tissues, and are non-Hookean in their behaviour. Most biological materials are anisotropic.

A variety of examples is shown in Fig. 2. Viscoelasticity, time-dependence, stress-relaxation, hysteresis and so on are exhibited by many biological materials as shown in the examples of Fig. 3, and the properties are affected by the environment—particularly by water content. It is a feature of biological materials that the values of these properties depend on which part of the animal or plant the specimen is taken from (the materials are nonhomogeneous in themselves and differ from body to body) and on the history of the specimen. Significant variations in supposedly the same property thereby arise.

Experiments with biological materials are not always easy to perform: it is often extremely difficult both to make and to grip biological test pieces. Recently we have employed a microtome, instrumented to measure cutting

A

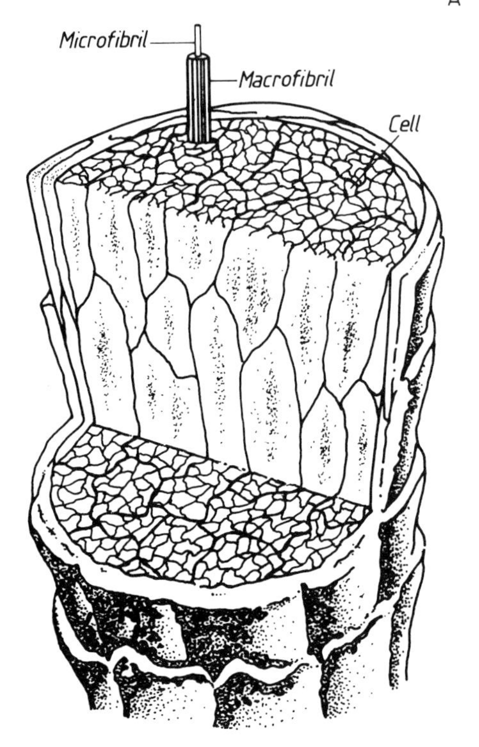
Microfibril
Macrofibril
Cell

B

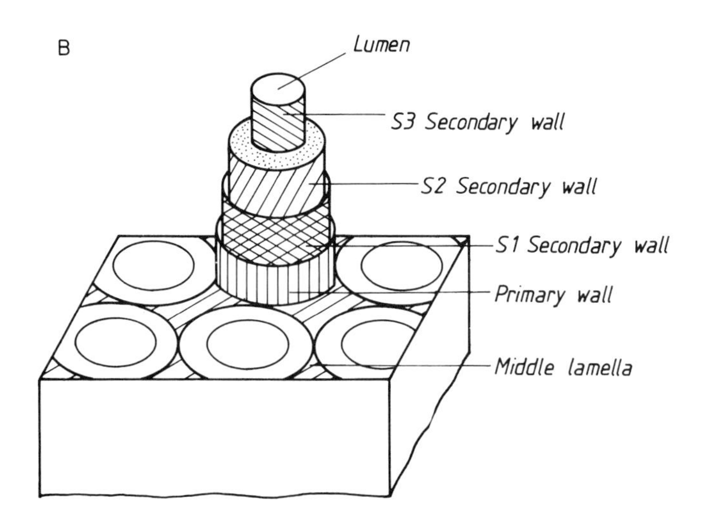
Lumen
S3 Secondary wall
S2 Secondary wall
S1 Secondary wall
Primary wall
Middle lamella

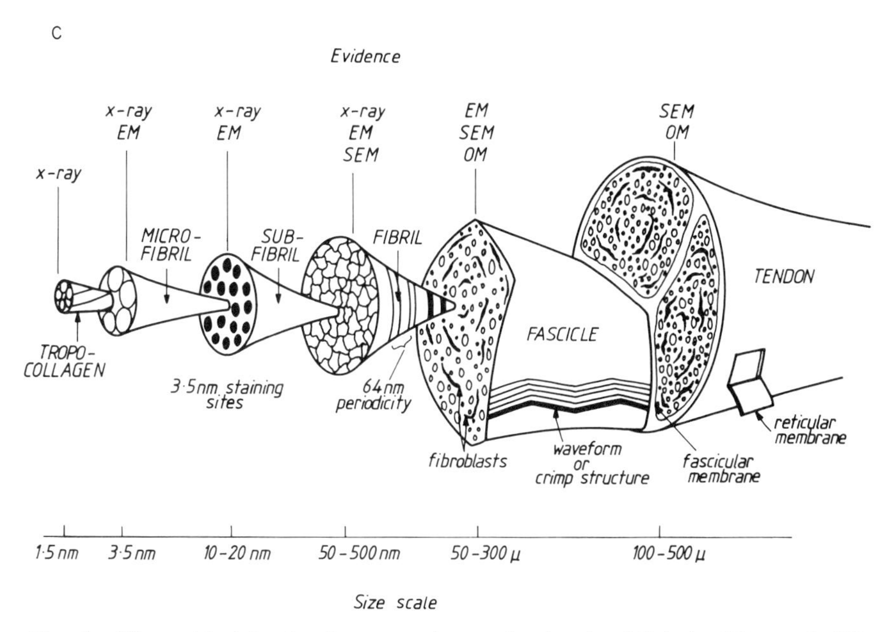

Fig. 1 Hierarchical levels of structural organization in: (*A*) hair; (*B*) wood; (*C*) tendon (after Kastelin and Baer 1980, SEB Symposium 34)

force (and thereby the fracture toughness) of biological materials, which overcomes many of these problems, Fig. 4*A*. A typical cutting force/blade displacement diagram is shown in Fig. 4*B*. It is noteworthy that the instrument picks up variations in properties across a slice (cortex-medulla-cortex in carrot, say, or growth rings in wood). The traces are highly reproducible on successive slices. The total work of cutting is given by the area under the cutting-force/blade-displacement diagram. The work comprises three components: friction, flow and fracture. An optimum relief angle between blade and cut surface exists at which the cutting force (and hence damage to the specimen) is least. The forces generated by cutting at angles smaller than the optimum are much affected by friction; at larger angles, the forces are higher owing to the greater deformation in the offcut and associated greater work of section curling (Fig. 4*C*). Friction and flow components must be subtracted from the total work in order to determine work of fracture. Friction may be approximately accounted for either by observing the force reading as the blade traverses the cut face on the return stroke after cutting a slice, or by noting the forces as the blade is taken in the forward direction over a surface which has just been cut, but without actually cutting. The friction is comparatively small ($<10\%$), particularly in

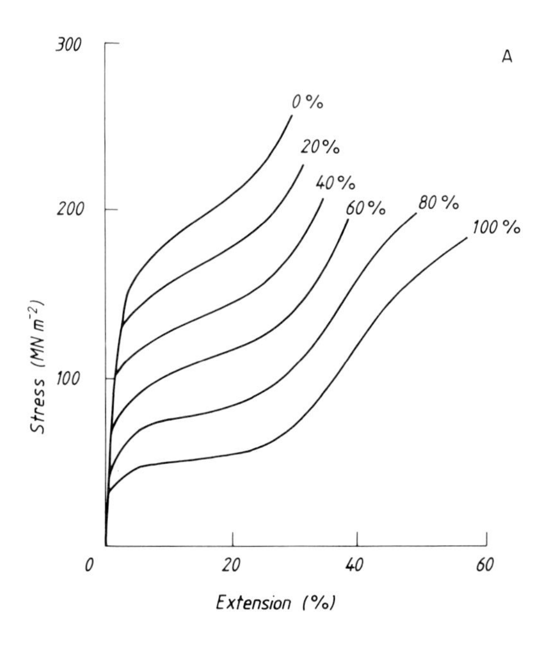
A
300
200
100
0
Stress (MN m⁻²)
0 %
20%
40%
60%
80%
100%
20
40
60
Extension (%)

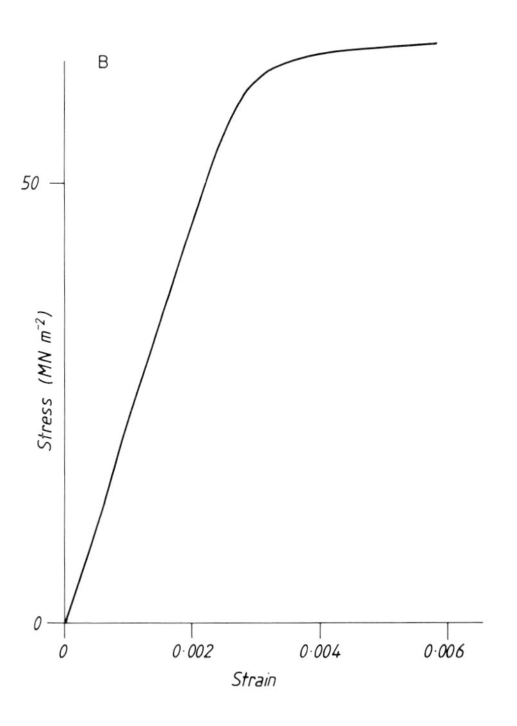
B
50
0
Stress (MN m⁻²)
0
0·002
0·004
0·006
Strain

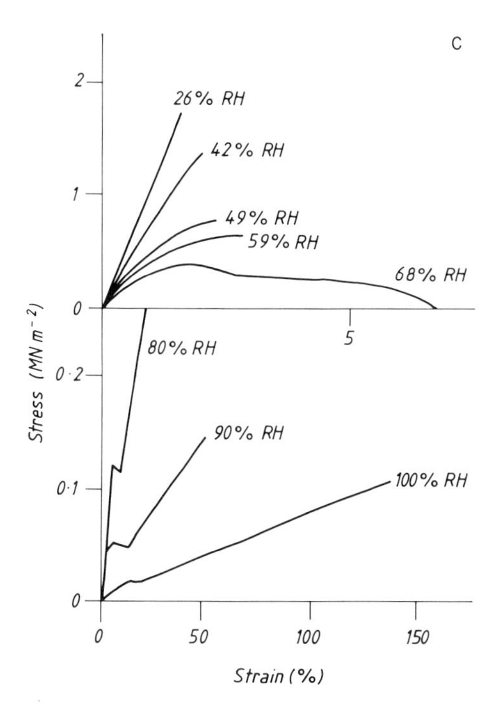

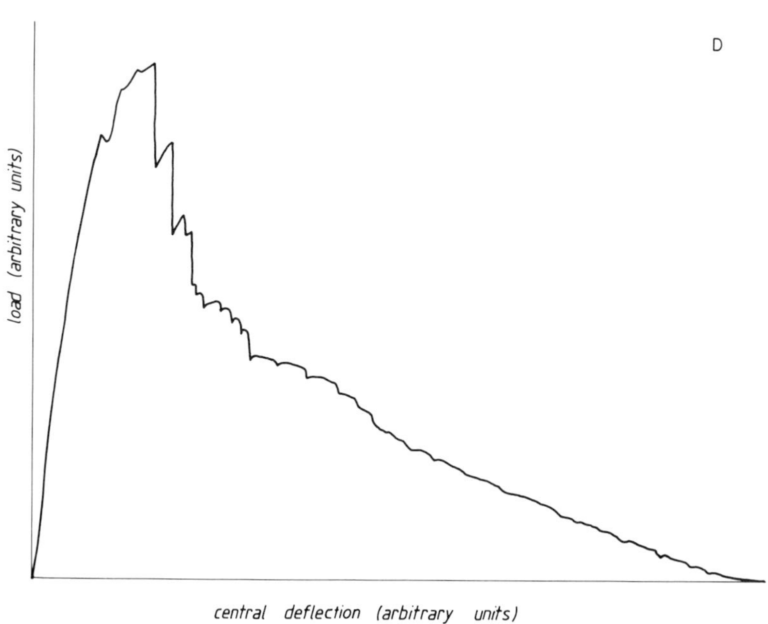

Fig. 2 Stress-strain curves for: (*A*) wool (α-keratin) at various relative humidities (after Peters and Woods, 1955; in SEB Symposium 34); (*B*) nacre (after Currey, 1980, SEB Symposium 34); (*C*) new-born rat skin (after Papir, Hsu and Wildnaner, 1975; in Structural Biomaterials); (*D*) sitka spruce in 3-point bending (after Jeronimidis, 1980, *Proc. Roy. Soc.* **B208**, 447)

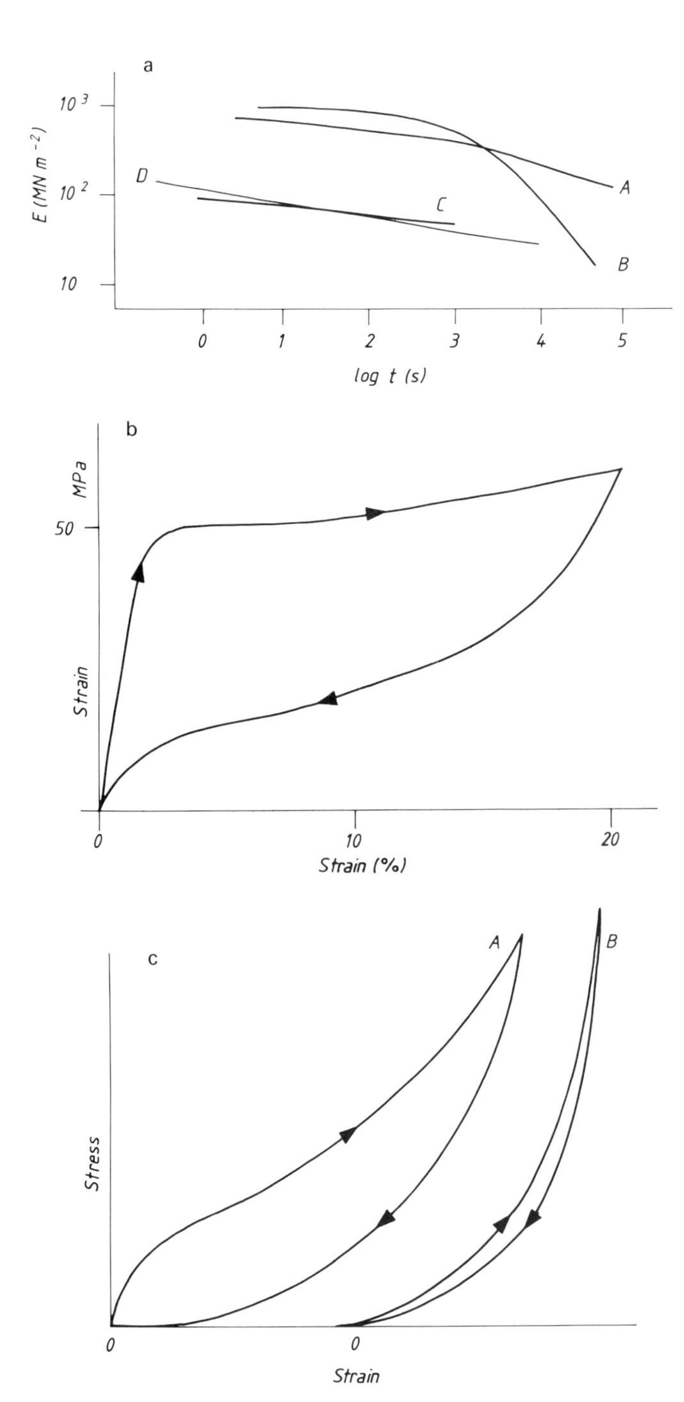
a
E (MN m −2)
10 3
10 2
10
D
C
A
B
0
1
2
3
4
5
log t (s)
b
MPa
50
Strain
0
10
20
Strain (%)
c
A
B
Stress
0
0
Strain

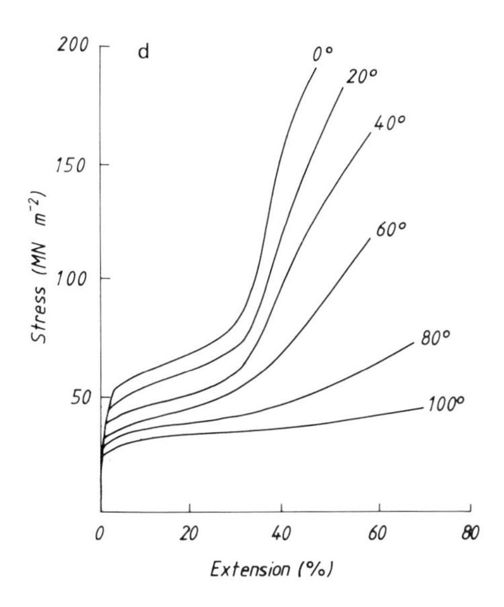

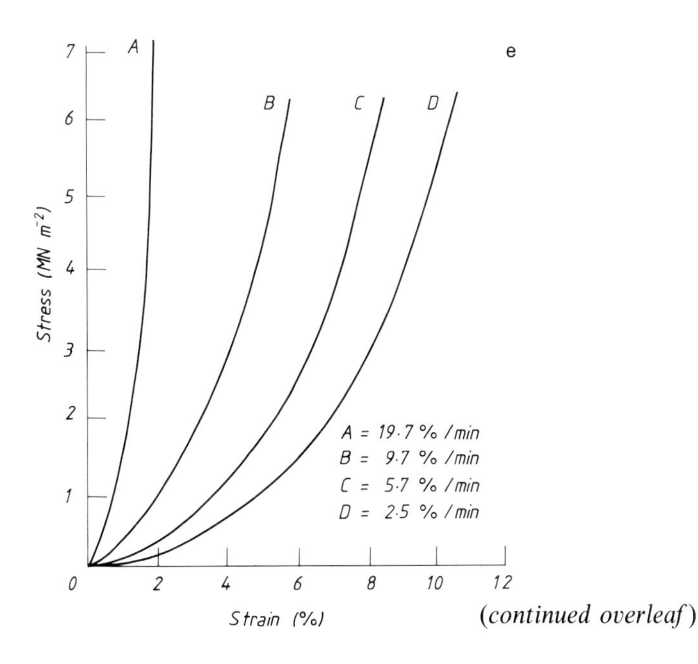

(continued overleaf)

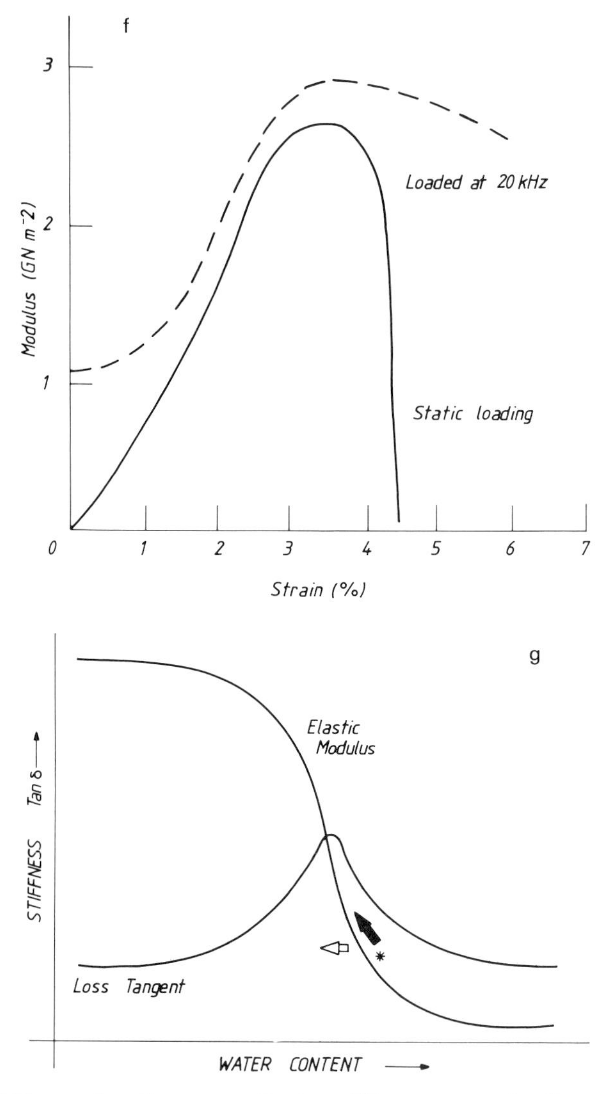

Fig. 3 (a) Stress-relaxation curves for two different types of collagen (A, rat-tail tendon at 3.5% strain; B, the same at 7.5% strain; C, mussel byssus thread at 5% strain) and polyethylene (D), a typical paracrystalline material (data of Wainwright 1965 and of Smeathers and Vincent 1979, Vincent Structural Biomaterials). (b) Hysteresis of α-keratin in hair (after Vincent). (c) Comparison of hysteresis cycles of rubber (A) and a typical collagenous biomaterial (B) (after Vincent). (d) Effect of temperature on the stress-strain behaviour of wool in water (after Peters and Woods, 1955; in SEB Symposium 34). (e) The initial part of 4 stress-strain curves of canine anterior cruciate ligament run at different strain rates (after Haut and Little, 1969). (f) The Young's modulus of rat-tail tendon as measured statically and at a very-high strain rate (20 kHz sonic modulus), (after Mason, 1965). (g) Stiffness and tan δ versus water content

biological materials containing water and, of course, small at angles of cut around the optimum or larger.

The data shown in Fig. 4*D* are for carrot cut at the otpimum angle (4°, as indicated in Fig. 4*C*) at a speed of some 3 mm . s^{-1} over the knife edge. The data level off at small thicknesses (about 225 μm for the cortex; 450 μm for the medulla) which are some three times the cell size in each instance. The corresponding toughnesses are 210 J . m^{-2} and 266 J . m^{-2}. There is no marked anisotropy in carrot, the data relating to longitudinal, radial and axial slices in both cortex and medulla. Flaccid carrot is tougher (around 300 J . m^{-2} depending upon the degree of flaccidity and the form of back-extrapolation used for the determination) and more rubbery.[4]

The instrumented microtome presents a remarkably easy method of determining toughness. Furthermore there is no need to measure the areas under the force/displacement curves for the plot against section thickness, as simple analysis shows that the toughness of a rectangular sample is given directly by the cutting force divided by the width of cut (Fig. 4*A*).

The general conclusions regarding mechanical properties of biological solids are that the results are pretty much what one would expect, i.e. in the case of the stiff materials, moduli of 10–100 GPa and strengths (depending upon the method of measurement) of 50–100 MPa. The exceptional results are for the fracture toughness of wood and antler, whose values of ca. 10 kJ . m^{-2} are an order of magnitude greater than they "should be". Again some shells have toughnesses of about 1 kJ . m^{-2}, even though the major ingredients (95%) are brittle material phases with hardly any crack resistance. Clearly there is something different about wood in the one case, and something special about the remaining 5% of material in shells. We shall discuss this later. In the case of extensible biological materials, the stiffnesses are orders of magnitude lower in the early stages of deformation, but where fibre orientation can take place considerable stiffening occurs at large strains. The values are then what one would expect on the basis of rule-of-mixtures composite theory, viz: in the case of skin and artery (with 40% and 20% volume fraction of collagen), final moduli of some 100 MPa. The fascinating thing about many extensible tissues is that although they are often difficult to tear, their fracture toughnesses are not all that large. There are lessons to be learnt from the reason why, which we shall see is connected with the shape of the stress-strain curves.

4 THE ROLE OF WATER IN BIOLOGICAL MATERIALS

Vincent[18] remarks that when the components of a biological material are listed, it is rare to find water included. For instance mammalian skin is

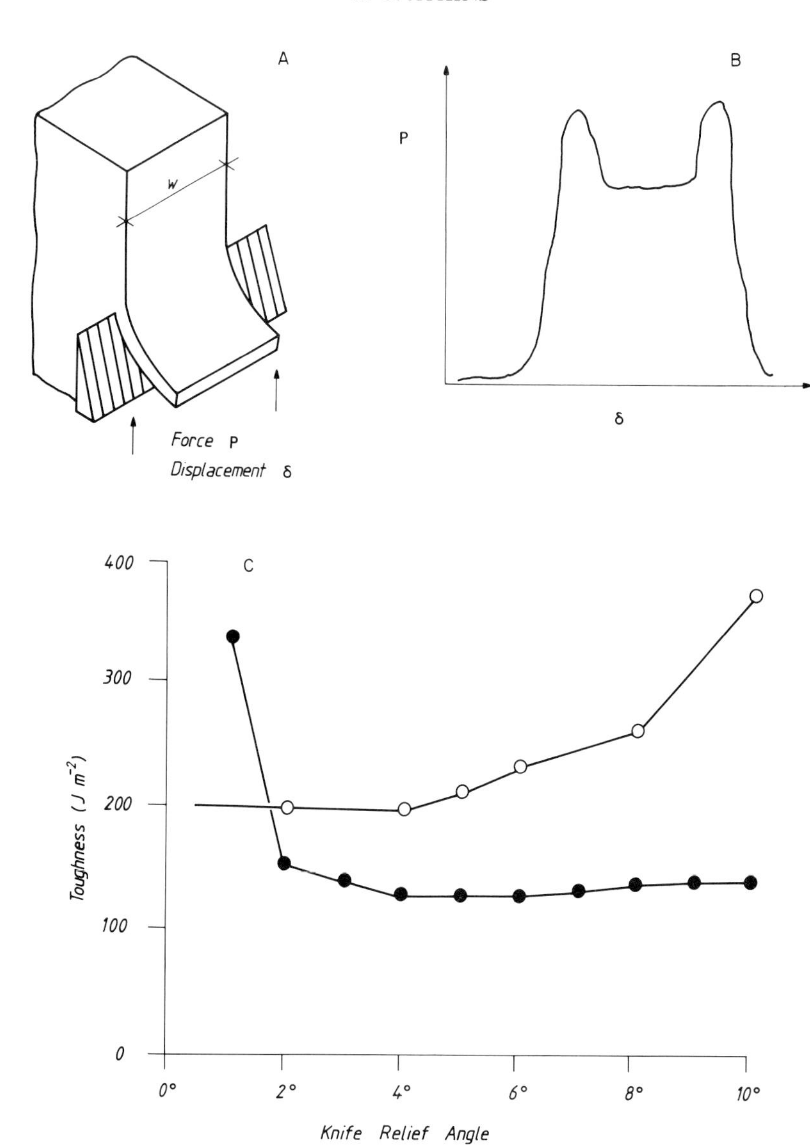

composed of collagen (the major fibre) and elastin (which contributes resilience) in a protein-polysaccharide matrix substance, the outer covering being dead cells of α-keratin full of a sulphur-containing protein. Water is not mentioned, yet without it skin becomes crinkly and brittle, as happens in the diseases of excema and psoriasis. Water acts as a plasticiser, dropping the glass transition temperature of most biological materials from about 200 °C

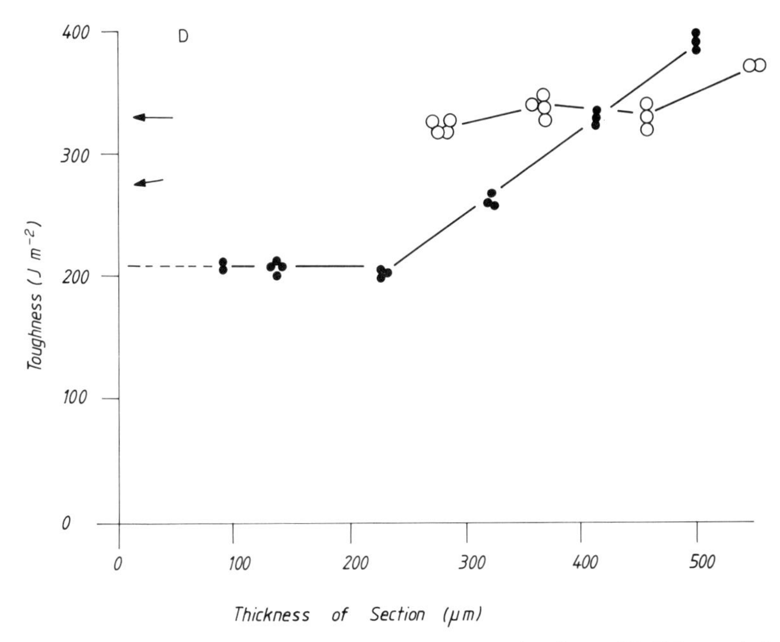

Fig. 4 (*A*) Microtome instrumented to measure cutting force. (*B*) Schematic force-displacement trace, showing different mechanical properties across a section. (*C*) Variation of microtome force (toughness of rectangular sample) with knife relief angle for histological wax (solid points) and carrot cortex (open points). (*D*) Variation of toughness of rectangular sample with thickness of section for turgid carrot cortex (solid points) and flaccid carrot cortex (open points)

to about $-10\,^{\circ}C$ or so (Fig. 5), without which they would be glassy. Water therefore must be considered a component of major importance, not only from a chemical viewpoint but also, as we shall see, from a mechanical viewpoint via the turgor pressure (cell pressure produced by osmotically active constituents, principally sugars).

Amino acids and sugars are polymerized in the aqueous environment of the living cell and its surroundings; side chains of the polymers interact with the water so that less polar groups clump together away from the water, thus directing both conformation and (at a later stage) aggregation. The polymer is thus fibrous or globular depending upon the type and order of the monomer units; this is hinted at in Fig. 6 which shows that the stiffer insect cuticles tend to be more hydrophobic. When the material is finally released from the cell into the extracellular environment, there to aggregate into a larger functional unit together with the secretions from other, similar cells, the cell still controls its water content, though whether directly or indirectly

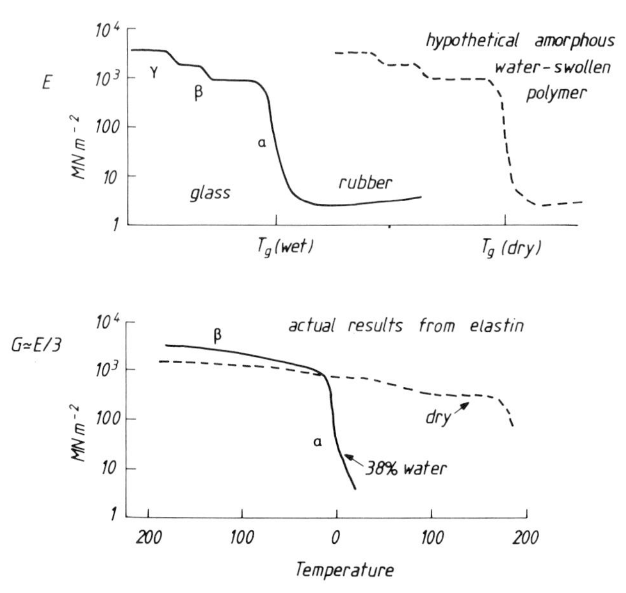

Fig. 5 Schematic viscoelastic transitions (after Vincent)

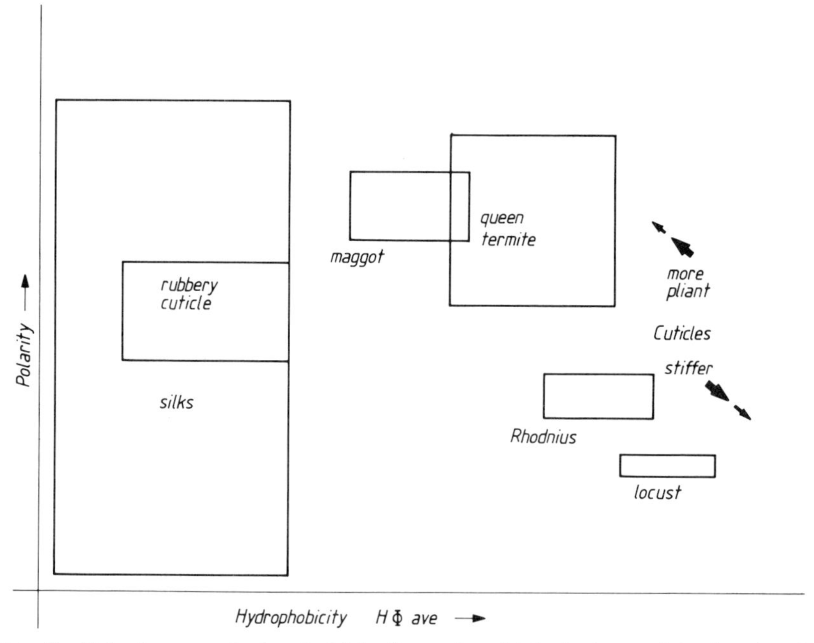

Fig. 6 Polarity versus hydrophobicity for various biological materials, showing that stiffer insect cuticles tend to be more hydrophobic (after Vincent)

(e.g. by changes in pH) is not known and will, in any case, vary with the example chosen. As a case in point, the blood-sucking bug, *Rhodnius prolixus* can change the stiffness of its cuticle (the outer covering of all insects) from 100 MPa with 25.8% water to 1 MPa with 31.3% water; it does this so that it can the more easily accommodate the huge blood meal which it takes once a month or so.[15] Again, the loss of elasticity of skin with age has been correlated with a drop in water content of about 3.5%[7] which affects the stiffness of the elastin. These large changes in stiffness with comparatively small changes in water content is a sure piece of evidence for a viscoelastic transition.

The influence of water content on the stiffness and fracture properties of grass leaves has been studied by Vincent[16] as it is important in a number of areas such as palatability and ease of grazing, resistance to trampling, ease of harvesting and handling, and response to drying as when making hay. In cross-section, the grass leaf appears as in Fig. 7*A*. The main fibrous tissue is the sclerenchyma, most of which occurs in distinct bundles of fibres, but some of which occurs in association with the bundles of vascular tissue which transport water and nutrients along the leaf. One surface is ridged, the other smooth. The (separate) volume fraction of both bundles and fibres is some 4%. Using composite mechanics, Vincent has shown that the tensile and fracture properties of grass laminae are governed by the fibres, which have a longitudinal modulus of some 10 GPa, the bundles displaying only 800 MPa. The stiffness of grass leaves changes with water content, increasing markedly below a water content of 20% of the dry weight (Fig. 7*B*) owing to the disappearance of "free" (i.e. liquid not bound) water which acts as a plasticizer. There is a similar transition in the stress-strain curves (Fig. 7*C*) where the two parts of the curve probably represent the properties of partially dried cells (the initial part) and the cuticle (final part) unfolded by stretching.

Fracture of grass leaves containing different length notches or cuts, shows that grass is *notch-insensitive* (Fig. 8), i.e. the strength is directly proportional to the load-bearing cross-sectional area; a notch-sensitive material on the other hand loses a very large proportion of its strength with only a small notch (fresh apples, potatoes and carrots are notch-sensitive). The notch-insensitivity of grass comes about because the matrix material binding the continuous parallel elements together does not transmit shear very effectively, so that if a single fibre is broken the stress is redistributed evenly among the other fibres (rather like the action in a rope) rather than being concentrated at any one point. Water content also affects the fracture behaviour of grass leaves. At high moisture contents, fractures pass straight across the laminae between the notches. Below 50% moisture content, however, the fracture takes a more complex course and is more readily deflected by the bundles.

Fracture *between* the fibres, parallel to them, becomes easier at lower moisture contents (when, at the same time, the longitudinal elastic stiffness of grass laminae doubles and the transverse stiffness increases about seven-fold). Work of fracture across the leaf veins is hardly changed with change of moisture (suggesting that the fibres are scarcely embrittled by drying) and even though the transverse stiffness of the cells between the fibres increases markedly with drying, grass remains notch-insensitive. That fracture between the fibres, parallel to them, becomes easier at low moisture contents

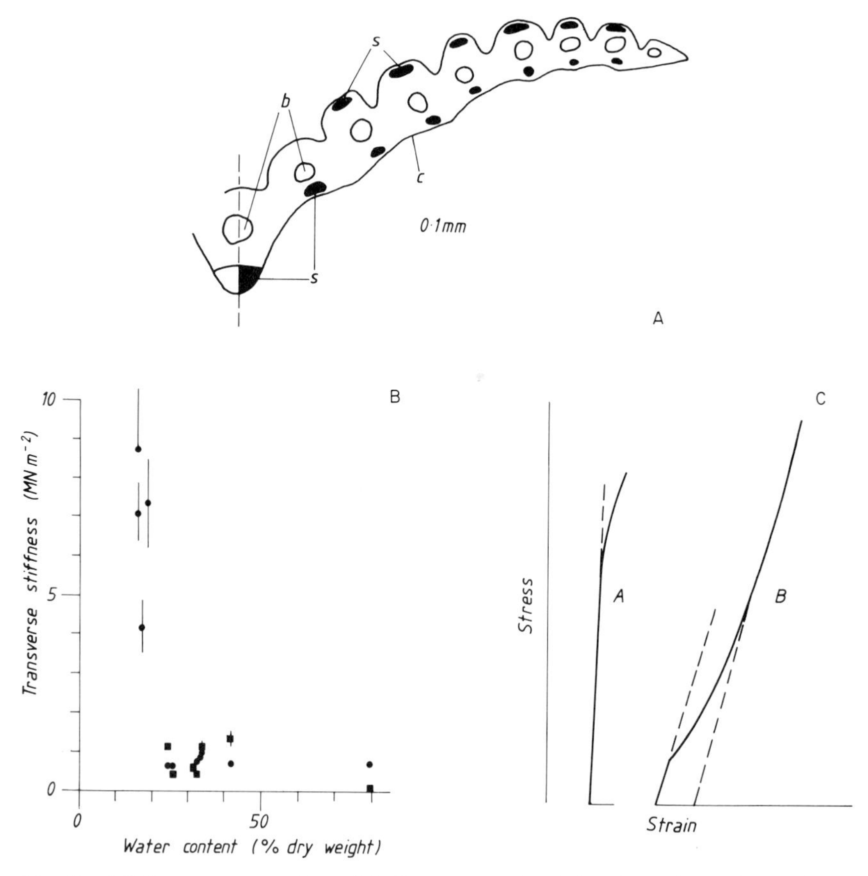

Fig. 7 (*A*) Transverse section of *Lolium perenne* leaf showing sclerenchyma fibres (s), vascular bundles (b), cuticle (c). The remainder is composed of relatively large thin-walled cells under turgor pressure which are considered as forming a homogenerous "matrix" phase. (*B*) Variation of transverse stiffness of grass laminae with changing water content. (*C*) Stress-strain curves showing the variation in transverse stiffness of grass laminae with (A) less or (B) more than 20% water. The dotted lines show the gradients used for estimating stiffness in Fig. 7*B* (after Vincent, 1983)

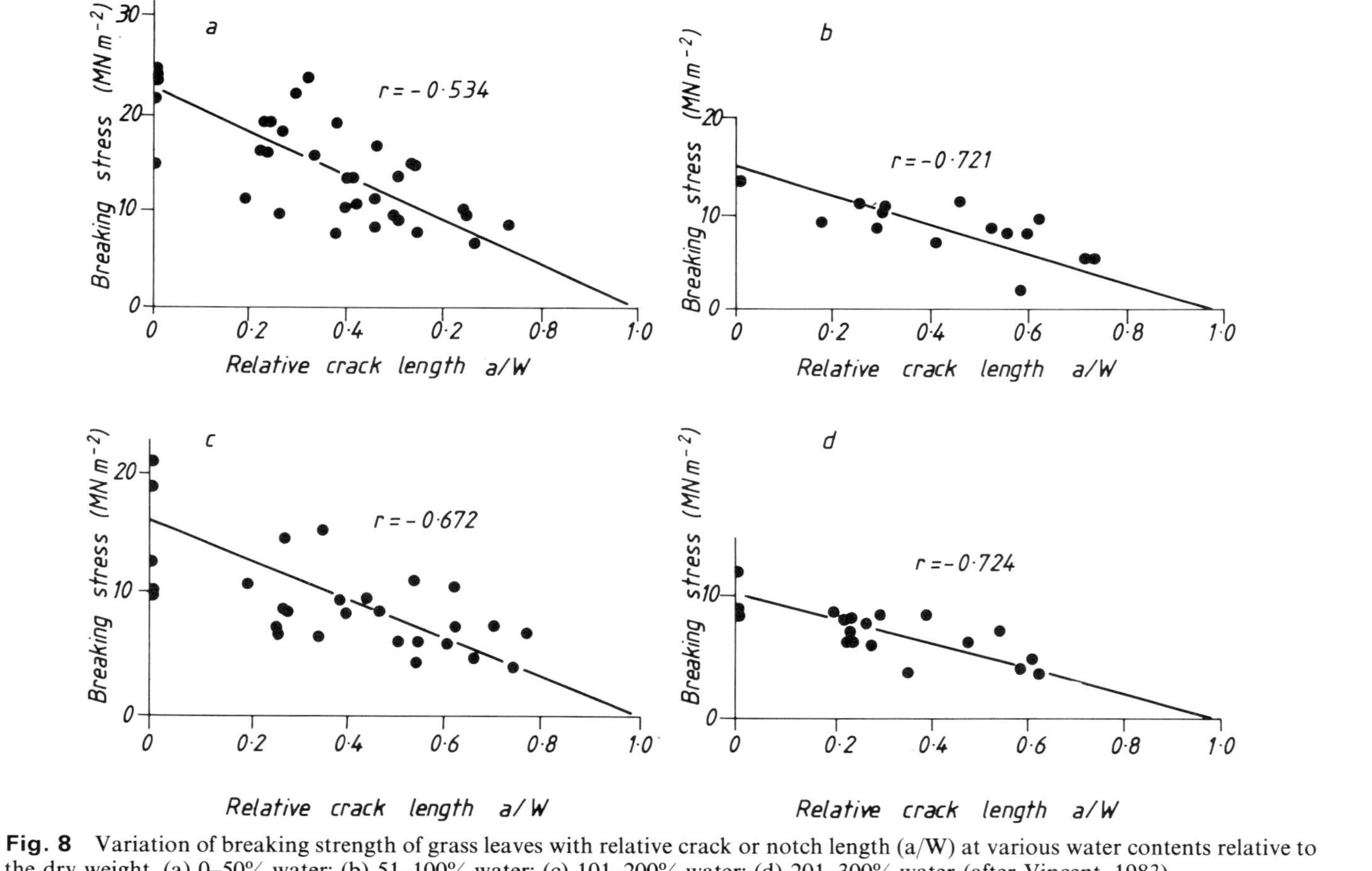

Fig. 8 Variation of breaking strength of grass leaves with relative crack or notch length (a/W) at various water contents relative to the dry weight. (a) 0–50% water; (b) 51–100% water; (c) 101–200% water; (d) 201–300% water (after Vincent, 1983)

presumably explains "hay-shatter" on drying, i.e. hay fractures when it has been allowed to dry out below the recommended water content of 25%,[14] and, as a consequence, up to 14% of the total dry matter can be lost. This clearly is of commercial importance and it would be of value to breed a shatter-resistant hay. As Vincent[16] points out, however, it is evident from the experiments reported here that increasing the fibre content of the grass lamina, making it "stronger", will not alleviate this problem. Similarly it is not apparent that manipulation of other variables of the lamina—morphological or biochemical—will necessarily have any effect on the fracture properties.

That water acts as a plasticizer for biological materials is perhaps not unexpected and the consequences are similar to well-known environmental, rate or temperature effects on solids commonly used in engineering. There is however an additional, and different, role played by fluids on mechanical properties via turgor pressure (i.e. pre-stressing of the cells of biological materials). This is particularly well-illustrated by some unpublished work on dandelion stems conducted by Jeronimidis and Vincent.[13]

If a dandelion stem is dipped into sugar solution its bending stiffness falls and it wilts which is usually explained in terms of a reduction in the turgor pressure. More detailed experiments show that the drop in bending stiffness EI with increasing sugar concentration takes place with hardly any change of tensile modulus E (Fig. 9*A*) where I is the second moment of area of the cross-section. If longitudinal strips are cut from dandelion stems (Fig. 9*B*), they take up the curved shape shown in Fig. 9*C*, the equilibrium radius of curvature of which can be measured; the *outside* of the stem is on the *inside* of the bent strip. Immersion in distilled water produces more curvature in the strips, whereas after immersion in sugar solutions the strips get more and more floppy and eventually become like bits of string, and lose all bending stiffness. Is it possible, however, to predict the radii of curvature of the strips from the concentration of the sugar solution, and to establish whether the drop in bending stiffness could be related to the diminution of turgor pressure?

The structure of a dandelion stem is shown schematically in Fig. 9*D*: the cells composing the structure extend axially along the stem like sausages and, significantly, have different cross-section sizes, being of large diameter and thin wall thickness on the *inside* of the stem with a progressive change to small diameter and thick cell walls on the *outside*. Consequently, the cell radius: wall thickness (r/t) ratio changes being about 15:1 on the inside but only 2:1 on the outside. The individual cells are pressure vessels, whose internal pressure is the turgor pressure of the plant. Turgor pressures of up to 20 atmospheres have been measured in plants, but some 10 atmospheres is more likely in this case. Hence using elementary thin-walled cylinder theory

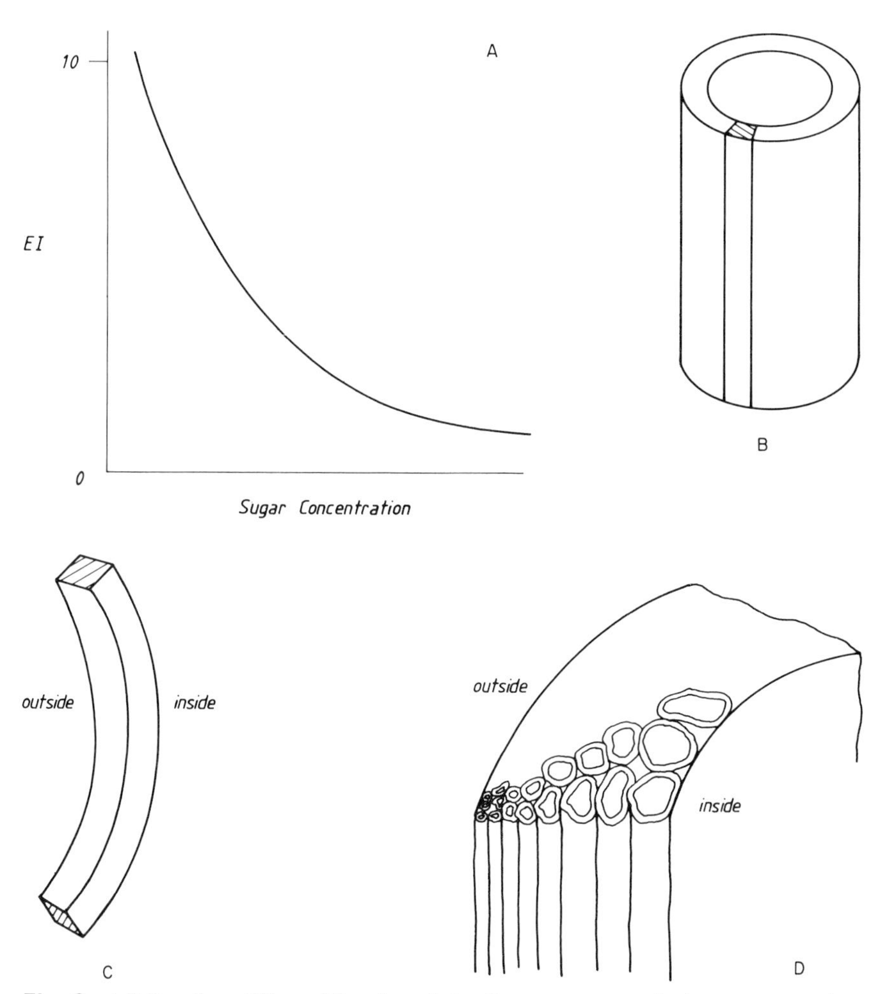

Fig. 9 (*A*) Bending stiffness *EI* and modulus *E* versus sugar solution concentration for dandelion stem; (*B*) longitudinal strips cut from wall of dandelion stem; (*C*) curvature of strips cut from wall of dandelion stem; (*D*) cell structure of dandelion stem (after Jeronimidis and Vincent)

the axial stress at the inside of the stem is some $\sigma_{\text{inside}} = pr/2t = (10/2)15 = 75$ atmospheres or about 8 MPa. Of more importance than the absolute value however is that the axial stresses differ at interior and exterior locations owing to the different r/t values (the turgor pressure is probably the same throughout the plant). Consequently the strains must also differ, being greater on the inside than outside. This explains why strips, cut from the stem, curl in the fashion they do.

The stress state of the cells in the complete stem of the free-standing plant will be different from that in the "released" strip and the problem may be attacked in the same way as residual stress problems are tackled in analyses of strengths of materials. In order to restore the curved strips to their original straight position in the stem, an applied bending moment M is required (Fig. 10). The combined stress state thereby produced is the (uniform) stress state of the cells in their constrained positions in a straight stem. Using elementary bending theory to link the curvature of the strip with the difference in turgor pressure as between cells inside the stem and those outside, together with the mean axial stress produced by the turgor pressure, Jeronimidis and Vincent have shown that the radius of curvature R falls hyperbolically with increasing turgor pressure. Hence strips cut from the stem of dandelion curve more when the turgor pressure increases as a consequence of immersion in distilled water, but curve less when the turgor pressure reduces on immersion in sugar solution (cf Fig. 9*A* and *B*). With appropriate values for the Young's modulus of the cell wall and other parameters, theory predicts $R \approx 4$ mm; experiment gives $5 \sim 8$ mm. Given the uncertainty of the magnitude of the turgor pressure, and practical problems associated with anticlastic curvature, the results are good.

It is of interest to calculate the stress in the cellulose fibres of the cell wall. We showed earlier that the mean wall stress on the inside of the stem is some 8 MPa. The local volume fraction of cellulose on the inside of the stem is about half the mean volume fraction of cellulose (i.e. about $0.07/2 = 0.035$), and this represents the *real* cross-sectional area taking the load rather than the apparent area given by the wall of the cylindrical cell. As the matrix

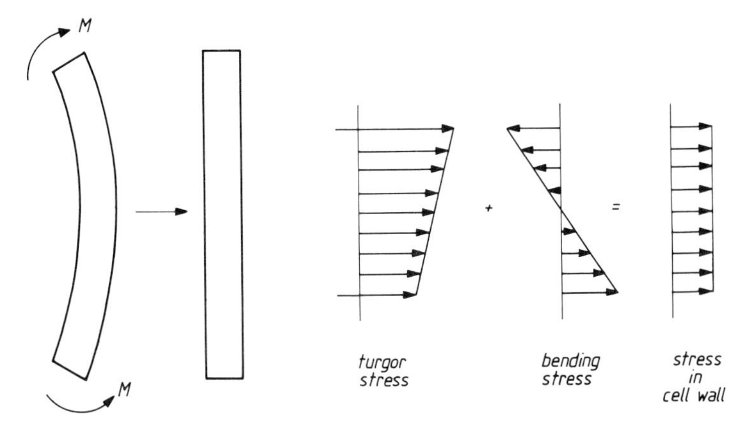

Fig. 10 Stress distributions in stem of curved and straight dandelion caused by turgor pressure (after Jeronimidis and Vincent)

material in the cell wall has negligible load-carrying capacity, the stress in the cellulose fibres must be some 8/0.035 = 230 MPa. In comparison, we remember that the yield strength of a typical steel is about 300 MPa.

5 THE "ENGINEERING" OF BIOLOGICAL MATERIALS

Since biochemistry dictates that natural systems are often fibrous, tensile forces are easily resisted (as in ropes). Unlike many manufacturers of artificial composite materials, Nature rarely provides strong interfacial adhesion between matrix and fibres. Isolation of the elements in this way limits the shear connexion between fibres so that if one filament breaks there is no path by which the released energy can reach other elements and cause them, in turn, to break in a "domino" fashion. Hence bodies, whose work of fracture is not really all that great, can be remarkably resistant to rupture. Examples of this sort of thing are mammalian tendons, spiders' webs and so on.

Animal membranes which resist multiaxial tensile loading, cannot have weak interfaces if they are to be watertight or gas tight. Yet they are usually remarkably difficult to tear (it is difficult to paunch a rabbit with a blunt knife). Furthermore, unlike balloons or pressurized metal structures, such animal membranes do not explode when pierced. One might presume that the fracture toughness must necessarily be high, but in fact it is typically $< 10\ \mathrm{kJ/m^2}$—which is an order of magnitude less than that of aluminium foil, say, which does tear very easily. Gordon[8] has pointed out that where rat skin, or worm cuticle or human arteries differ from metal sheet—or for that matter from rubber sheet—is not in the work of fracture but rather in the shape of the stress-strain curve, Fig. 11. Extensible biological materials are not Hookean elastic and have instead a J-shaped curve and this characteristic

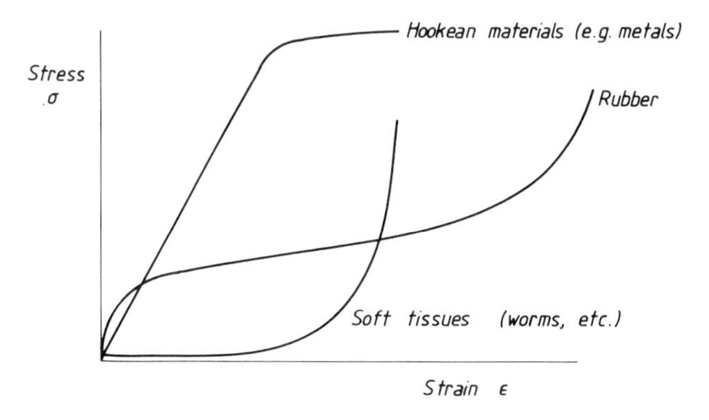

Fig. 11 Different characteristic shapes of stress-strain curves (not to scale)

shape plays a crucial role in fracture behaviour. The initial part of the curve is almost horizontal, rather like surface tension in a liquid, and hence the material has a very low shear modulus. There is thus virtually no communication of strain energy between parts of the stressed body and it is difficult to feed energy into potential crack sites. Such materials therefore do not tear; however, they may be cut, but only easily with sharp tools which concentrate the deformation into narrow zones. In contrast, unreinforced rubber has a sigmoid-shaped stress-strain curve with a high shear modulus at low stresses. Intercommunication between elements of a rubber sheet is good and rubber is easily torn, even though its work of fracture is comparatively higher. A J-shaped curve is produced when knitted fabric is pulled in tension, the progressively increasing stiffness at larger strains arising from progressive orientation of the fibres towards the pulling axis. Knitted fabrics are difficult to fracture and, in a related fashion, we note that tears in cloth are L-shaped along warp and weft, and never along the 45° direction which displays a J-curve and into which it is therefore difficult to transfer energy. Progressive orientation of fibres is the reason for J-curves in biological materials, e.g. X-ray diffraction studies show that randomly-orientated collagen fibres in the wall of pig aorta progressively line up with increasing deformation. Only in situations where membrane fracture is intended (amniotic membranes, egg shells) does nature provide a quasi-Hookean stress-strain behaviour, otherwise J-curves predominate.

Living organisms require materials to resist compression and bending as well as tension. Nature's solutions to the problems of withstanding compressive and bending forces in systems made up of fibres depend on the particular mode of failure which is being guarded against, which in turn depends upon the following factors:[11]

a the loads and the distances over which they have to be carried
b the composite nature of biological materials and their mechanical anisotropy.

In compression, the possible failure modes are (a) Euler buckling; (b) crushing or shearing; (c) local buckling, either of cells or of fibres.

In bending, the possible failure modes are (a) tensile failure; (b) compressive failure, (by crushing, shearing or local buckling); (c) interlaminar failure in composites.

Most biological structures operate at low structural loading coefficients (i.e. the stresses are low compared with the "ultimate" stress, and the distances are greater than the lateral dimensions of the elements). Under these conditions the design criteria reduce to (a) controlling deflections in bending and (b) controlling Euler buckling in compression. In both cases the important parameters are the bending stiffness and the resistance to bending

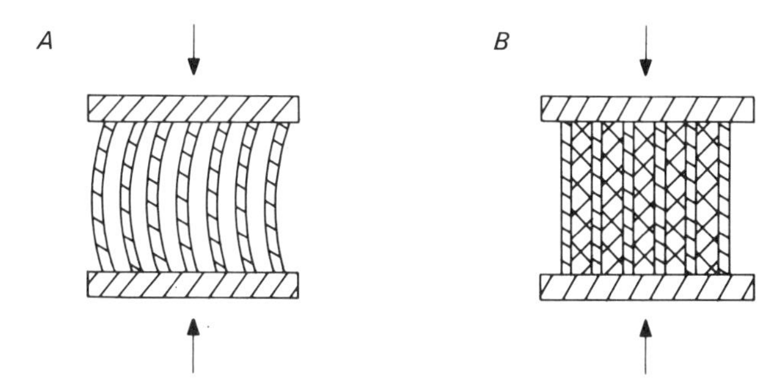

Fig. 12 Compressive behaviour of a sub-divided structure. (*A*) Separate elements buckle separately at a low load (e.g. grass and doormats); (*B*) compressive strength is much improved by some lateral communication (e.g. timber), (after Gordon)

failure both of which, when dealing with composite materials, depend on the fibres, the matrix and the degree of intercommunication. Note that a large EI does not necessarily require a high Young's modulus E; judicious distribution of material may give a large enough second moment of area I even if E is comparatively small. Furthermore, for fixed mass (in variable density structures), maximum I is given by bodies with least density ρ.

Three principal methods of designing against compressive and bending loads are found in Nature, (a) cellular structures (as in wood); (b) the use of turgor pressure (as in the dandelion and at the tips of growing plants, other parts of which have been lignified); (c) the addition of minerals (as in bone and shells). As a rule, some lateral communication between elements is necessary in compression structures in which case the possibility of fracture is more likely, Fig. 12. Nature is very careful to control the interfacial adhesion however and most fibrous compression systems are provided with "weak interfaces" which play a crucial role in determining the overall mechanical properties of wood, teeth and so on. The Cook-Gordon effect often comes into play (Fig. 13), whereby a weak interface in the path of a running crack will open up, slow down the crack and even divert it, thereby improving resistance to fracture. The role of interfaces is well-illustrated by Currey's discovery of the comparatively high work of fracture of mother-of-pearl (the material of mussel shells). It consists of 95% chalk (which has negligible fracture resistance) yet the shell material as a whole has a respectable fracture toughness of about 1 kJ/m^2 owing to the special role played by the remaining 5% interface material which is a mixture of protein and polysaccharide. Again, the cuticles of beetles are made from crossed layers of chitin fibres, rather badly stuck together by means of the resin-like substance called sclerotin. In bending, total isolation of elements is possible (rather like the tiles on a roof or the plank decking of simple wooden bridges); as Gordon

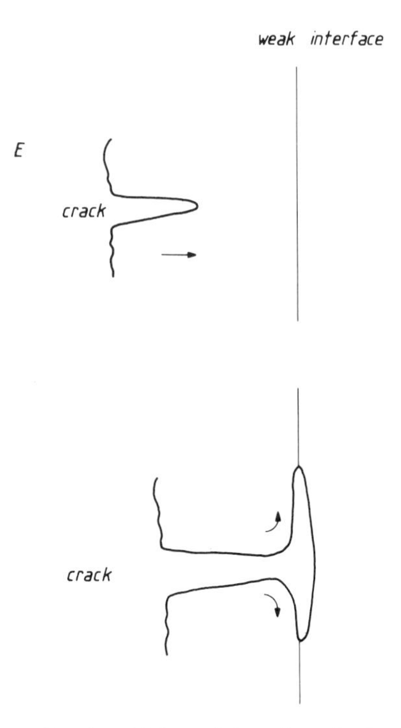

Fig. 13 Cook-Gordon effect in which weak interfaces in the path of a crack open up and thereby slow down and possibly divert the crack, thus increasing the resistance to fracture

remarked,[8] the aerodynamic surfaces of birds are not constructed from aluminium plates but from many isolated cantilevers of keratin (i.e. feathers)—and neither the individual barbs nor the feathers themselves are connected in any strong fashion.

An exciting realization by Jeronimidis and Gordon[12] was that wood is tougher than expected. The tubular cell walls of wood themselves have a hierarchical structure of spirally-wound cellulose fibrillae embedded in a matrix of lignin, Fig. 14. Eighty per cent of the material in a wood cell consists of the so-called secondary layer (S_2) which therefore controls the load-bearing capacity. The fracture toughness of most softwoods and hardwoods ($\sim 10\ \mathrm{kJ/m^2}$) is exceptionally high in relation to other mechanical properties. This cannot be explained on the basis of the "pull-out" contribution to toughness (which predicts at most $1\ \mathrm{kJ/m^2}$ for wood)—indeed wood fracture surfaces do not have an excessively stubble-like appearance. The "substructure" of biological materials is vital in this case for it is the helical

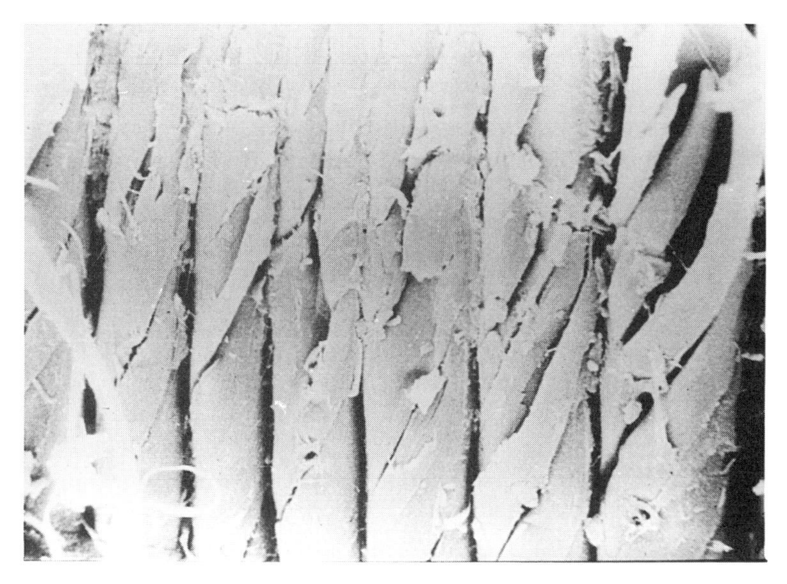

Fig. 14 Tensile decohesion and buckling of wood cells, giving large extensions and high works of fracture

arrangement of the cellulose fibres in the load-bearing S_2 walls that provides the clue. When pulled in tension the fibres decohere, and the cell walls buckle inwards into the lumen in such a way that longitudinal extensions of the cell up to 20% are possible. Thus, although the elastic fracture strain in wood as a whole seldom much exceeds 1%, cells close to the fracture surface extend far more and absorb much energy as they do so. Arrays of model cellulose fibres, made by winding glass and carbon filaments into hollow helices with resin, exhibit experimental toughnesses up to 400 kJ/m^2, which is much improved over the usual level of toughness in man-made composites. The orientation of fibres in the S_2 wall is a compromise between stiffness and toughness. In grass the fibres are more parallel to the cell axis than in wood, so the cell is stiffer but less tough; even so grass when trodden on buckles, deflects and eventually recovers.

Having discovered the ways in which Nature tackles the problems of strength, stiffness and toughness, can these methods be applied to engineering structures? This topic has been reviewed elsewhere[2,3] and will not be repeated here. Suffice to say that the Gordon-Jeronimidis helical tube solution for simultaneous high toughness, high strength and high stiffness, for example, looks extremely promising—particularly since holes are cheaper than carbon fibres!

6 APPLICATION TO AGRICULTURE AND FOOD PROCESSING

A better understanding of the mechanical properties of biological materials in themselves suggests application on a number of fronts. The commercially-important problem of hay-shatter has already been discussed in Section 4. A useful direction for investigation to develop a shatter-resistant hay may be to vary the relative amounts of water-binding components. It is not clear how the fracture properties will vary; in general cellulose will bind less water than other components of the grass lamina. In this connexion, the mechanical properties of grass account for the way different animals graze. If grass laminae are notch-insensitive then there is nothing to be gained by biting into it and hoping to propagate the crack so formed. The notch made by the teeth will not function as concentrator of stress and so will not make the grass easier to break. So if the animal is big enough (as is a cow) it breaks the leaves using the brute strength and no teeth. The method used by sheep, horses and geese is somewhat different and, according to unpublished studies by Bignall and Vincent, probably involves the introduction of a compression crease into the lamina by pulling it through a sharp angle. This crease then acts as a notch simultaneously introduced into all the sclerenchyma bundles on one surface of the leaf, depending upon which surface was on the inside of the angle and received the damage, and so makes fracture easier. It is clear though that this form of damage can be introduced into only a few laminae at a time: the geometry of a large bundle of grass leaves precludes the sharp angles necessary for the introduction of a compression crease in all leaves.

The mechanics of field crops may be changed without recourse to normal genetic selection. For example, the John Innes Institute is able to vary parts of plants independent of other parts without affecting the yield. But what are desirable changes? New information on mechanical properties may give the answer for problems such as crop-lodging. Again, plant breeders can breed for specific improved properties which are desirable. Rather than planting whole fields, a few plants can be grown instead and properties established by relatively simple laboratory tests (or even simple tests in the field). One example relates to the splitting of apple skins before harvesting. Figure 15 provides a comparison of the stress-strain behaviour of the skins of Golden Delicious and Cox's Orange Pippin apples. Golden Delicious variety, for example, possesses the same stiffness in transverse and longitudinal directions, and moreover the properties are the same in the young growing apple as in the mature apple. While the behaviour of skin from mature Cox's apples was also isotropic with also similar values of stiffness as Golden Delicious, there was a marked difference in the behaviour of young Cox's apples. The transverse stiffness was less than half the longitudinal stiffness and this

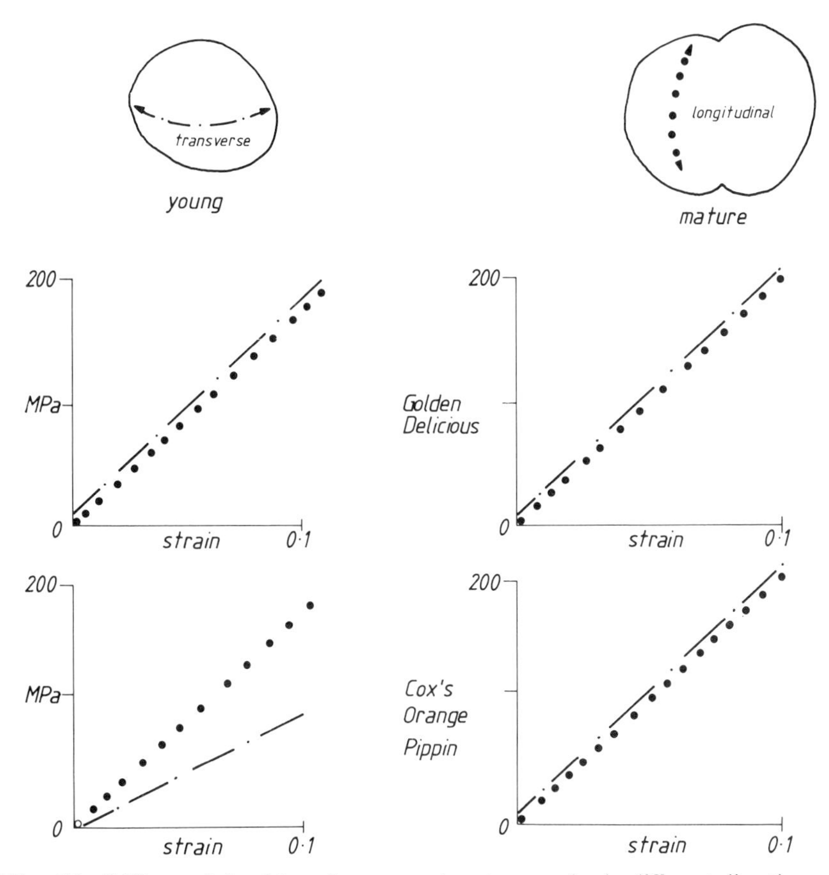

Fig. 15 Stiffness of the skins of young and mature apples in different directions

accords with the growth pattern of this variety in that young Cox's apples grow circumferentially much more than longitudinally. However, Cox's Orange Pippin apples suffer a great deal from splitting especially if rain follows a dry period. The skin has dried and "hardened" and splits when the apple increases its water content.

An understanding of the mechanics of cutting now enables histological sections more easily to be microtomed to higher quality. Related to this is the design of food processing machinery from an understanding of the mechanical properties of foodstuffs. Topics as diverse as meat slicing and grain milling are involved. Many fascinating questions may be answered by knowledge of mechanical properties of foodstuffs—what, scientifically, is "tough" meat? What is perceived "texture" in foodstuffs? There are many more.

Acknowledgments

I have drawn heavily on the work of others in the preparation of this paper, in particular my colleagues Professor J. E. Gordon, Dr G. Jeronimidis and Dr J. A. V. Vincent of the joint Engineering/Zoology Biomaterials group in Reading who have all been extremely helpful and who have permitted me to quote extensively from their work. In turn, we all acknowledge the activities of our students over the past years. Debt is also owed to Professor J. D. Currey and other workers with whom the Reading group collaborates for advice and guidance.

References

1. Atkins, A. G. (1982). The mechanics of microtoming, SEB Symposium, Leiden.
2. Atkins, A. G. (1983). Biomaterials in Industry, British Association for the Advancement of Science Annual Meeting.
3. Atkins, A. G. (1984). *Spectrum COI* (in press).
4. Atkins, A. G. and Vincent, J. F. V. (1984). *J. Mater. Sci. Letters* **3**, 310.
5. Bignall, M. and Vincent, J. F. V. (1983). Unpublished research, University of Reading.
6. Currey, J. D. (1983). Biomaterials in Industry, Course Notes, University of Reading.
7. Dorrington, K. L. (1981). *The Lancet* **8214**, 264–265.
8. Gordon, J. E. (1979). NDT Conference, Monterey, California.
9. Gordon, J. E. (1968). The New Science of Strong Materials—or 'Why You Don't Fall through the Floor', Pelican, London.
10. Gordon, J. E. (1978). Structures—or Why Things Don't Fall Down, Pelican, London.
11. Jeronimidis, G. (1983). Biomaterials in Industry, Course Notes, University of Reading. (*Materials Science Club Bulletin No. 67.*)
12. Jeronimidis, G. and Gordon, J. E. (1974). *Nature* **252**, 116.
13. Jeronimidis, G. and Vincent, J. F. V. (1983). Unpublished research, University of Reading.
14. Jones, L. (1979). *Grass and Forage Sci.* **34**, 139–144.
15. Reynolds, S. E. (1975). *J. Exp. Biol.* **62**, 81–98.
16. Vincent, J. F. V. (1982). Structural Biomaterials, Macmillan, Basingstoke.
17. Vincent, J. F. V. (1982). *J. Materials Sci.* **17**, 856–860.
18. Vincent, J. F. V. (1983). Biomaterials in Industry, Course Notes, University of Reading.
19. Vincent, J. F. V. and Currey, J. D. (Eds) (1980). *Symp. Soc. Exp. Biol.* **34**.

9 The Fracture Behaviour of Meat—A Case Study

P. P. PURSLOW

A.F.R.C. Institute of Food Research—Bristol Laboratory

1 INTRODUCTION

The fracture properties of cooked meat are of fundamental relevance to its textural quality as the process of chewing meat is essentially one of breaking down the material in the mouth. In the field of industrial meat processing a thorough knowledge of the fracture behaviour of meat is also important, as many processes involve size reduction or fracture, e.g. flaking, dicing, guillotining, band-sawing, slicing and mincing, and the textural quality of the final food product can strongly depend on the quality of these fracture processes.

It is therefore important to have a fundamental understanding of the fracture behaviour of meat and how this relates to the structure of the material. The long-term aim of this is to explain and predict variations in the perceived texture of meat on the basis of variations in composition and structure, and hopefully to be able to control and optimize texture.

2 MUSCLE STRUCTURE

Muscle tissue is a highly complex structure that can best be described as hierarchical, i.e. each large subunit of an individual muscle is in turn an

FOOD STRUCTURE AND BEHAVIOUR
ISBN 0-12-104230-8

aggregate of smaller subunits, which in turn are aggregates of yet smaller aggregates. The most relevant levels of organization of muscle tissue in relation to this discussion are those shown in Fig. 1.

Figure 1*A* shows the left half of a whole fusiform muscle, with its tendon attachment running away to the left. It is surrounded by a thick layer of collagenous connective tissue, the epimysium. The whole muscle is made up of many muscle fibre bundles running along its length and is shown cut in cross section. A single muscle fibre bundle is shown in Fig. 1*B*. It, in turn, is made up of many individual muscle fibres. The fibre bundle has its own surrounding envelope of connective tissue, the perimysium. The perimysia of adjacent bundles merge to form a gross network structure internal to the whole muscle. An individual muscle fibre, shown in Fig. 1*C*, also has its surrounding envelope of connective tissue, the endomysium. The muscle fibre contains many myofibrils, which are the contractile elements of the tissue. The three levels of connective tissue are in themselves composite structures;

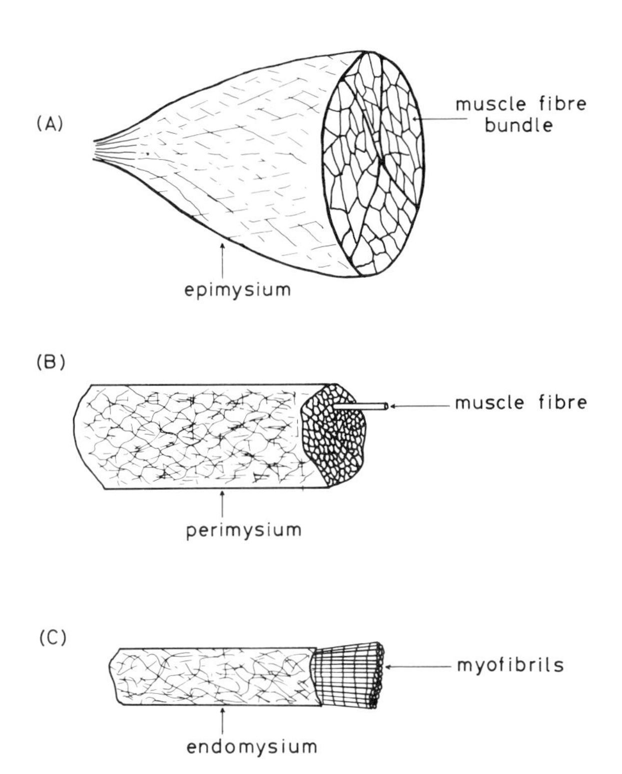

Fig. 1 The hierarchical structure of muscle. (*A*) Whole muscle. (*B*) A muscle fibre bundle. (*C*) A single muscle fibre

they are networks of collagen fibres in an amorphous ground substance. Figure 2 shows such a network of crimped collagen fibres in the perimysium of a beef muscle, *Cutaneous trunci*. The size of collagen fibres, their arrangement and the genetic forms of collagen present vary between epi-, peri- and endomysia.[1,35]

The transition of living muscle tissue to meat *post mortem* is essentially the development of *rigor mortis*, which involves the irreversible formation between the actin and myosin molecules in the myofibrils of the cross-links that in life form the basis of contraction. *Rigor mortis* is then "resolved" or reduced, as the meat is left to hang, by autolytic, enzymic degradation of the myofibrillar structure. The development and subsequent resolution of *rigor mortis* are strongly affected by conditions such as temperature and pH, and are known to affect the perceived texture of the meat. In view of the structural complexity of the material and the important effects of post-mortem conditions, it is perhaps not surprising that variability in meat toughness is not well understood.

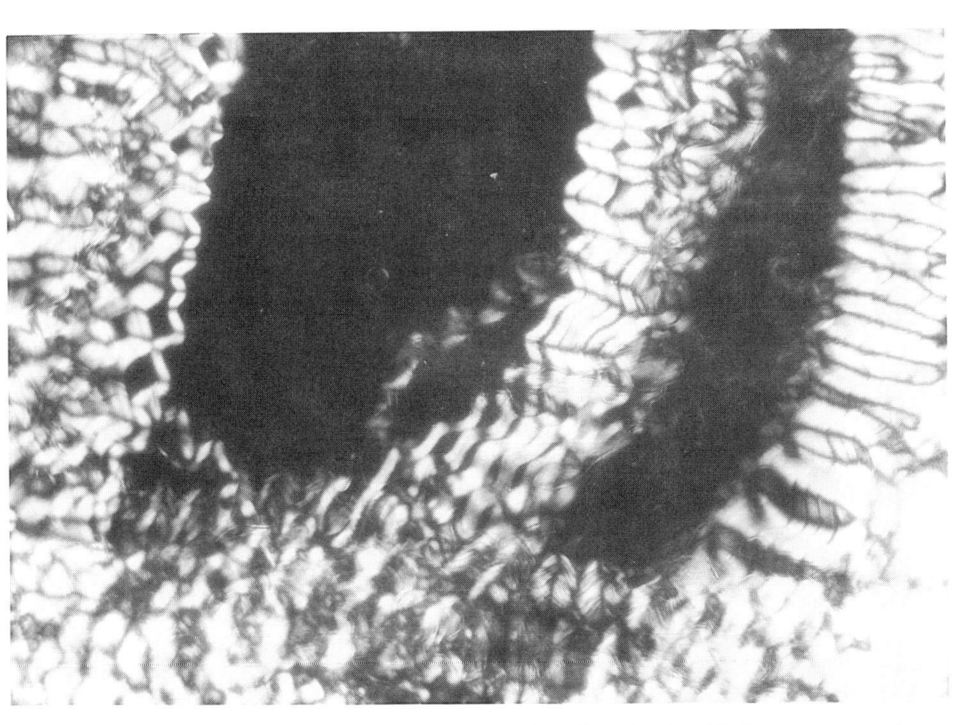

Fig. 2 Polarized light micrograph of perimysium from bovine *M. Cutaneous trunci*, showing orthogonal network of crimped collagen fibres

3 A BRIEF HISTORY OF TEXTURAL MEASUREMENTS

Researchers working on the sensory evaluation of texture have made considerable efforts in the past to develop objective mechanical tests to measure some aspect of the fracture process which is a good predictor of the perceived textural quality, or toughness, of cooked meat. Szczesniak and Torgeson[40] and Voisey[41] provide extensive reviews of the considerable number of instruments reported in the literature. These have largely been empirical test methods, the sole criterion of their value being the degree to which results correlate with taste panel scores. The most widely used and accepted of these are the shear presses, such as the Warner-Bratzler shear device;[6,7,44] and the Kramer shear press,[22,23] where one or more blades press or cut through a sample simply supported as a beam over the blade running slot(s); and the closely allied bite tests, such as the Volodkevich device[43] and the Winkler device[45] from which the MIRINZ tenderometer is derived.[26] These bite tests employ two axially opposed blunt wedges between which the sample of meat is squeezed and cut. All these techniques purport to measure the ultimate properties of meat in shear; measured peak force values are uniformly referred to as shear force values. However, as pointed out in the literature previously, the mode of deformation in these tests is anything but pure shear, being instead a complex mixture of compression, shear and tension, with the ultimate tensile properties being the major influence on the point of rupture.[29,41,42]

The Warner-Bratzler shear press seems to be the most widely accepted technique for measuring meat toughness,[40,41] to the extent that it is often used as a basis of comparison for other techniques[4,39] despite the fact that correlations between Warner-Bratzler shear values and taste panel scores have been inconsistent (see ref 40 Table XII).

The shear press and bite tests are somewhat unsatisfactory measures of the fracture properties of cooked meat for two reasons:

1 From a practical point of view, the highly empirical nature of these tests means that results obtained owe as much to the test conditions and apparatus as to the intrinsic mechanical properties of the material. The following have been identified as operational conditions affecting the results given by the Warner-Bratzler test:

Blade thickness[41]
Blade sharpness[36]
Angle of the blade edge[42]
Clearance between blade running slot(s)[41]
Rate of deformation[42]
Size of sample[29]

Some similar considerations also apply to the Volodkevich bite test.[11,41] Without strict standardization of these techniques, this means that comparisons between toughness values obtained in different laboratories using different apparatus and conditions are very difficult to make. Such a standardization, although called for in previous literature[41] does not appear to have been forthcoming, the closest approach being general guide-lines for a group of procedures to measure meat texture from an EEC working group.[2]

2 The second objection to the shear press and bite test methods is scientifically more important. Namely: (a) the highly complex pattern of stresses and strains set up in the shear press and bite tests;[41] (b) the way in which various components of the mechanical behaviour of the sample come into play during the test (i.e. shear and compression early in the penetration, tension in the final stages[29,41]) and (c) again, the dependence of loading patterns and test results on the conditions of test (as discussed in 1 above), mean that these tests do not measure any single fundamental mechanical property of the sample.

One of the aims of work on the mechanical properties of meat is to understand how the basic physical properties of the structure are determined by its composition and morphology, and which of these properties are important in the sensory evaluation of meat texture. Any such fundamental study is not aided by the use of shear press and bite tests as measures of mechanical behaviour because of their inability to measure fundamental mechanical properties.

In view of this, and because tensile properties dominate the mechanical behaviour of cooked meat at breaking point, simple tensile tests would seem to be a good starting point for fundamental measurements. Several workers have previously reported the use of tensile testing of simple strips.[3–5,11,24,27,30,37,38] Some of these have examined tensile properties before failure, such as extensibility[24,27] or initial yield point.[25] Stanley *et al.*[39] showed that ultimate tensile strength, UTS (i.e. stress at rupture of the material) is a good predictor of sensory value judgements on meat tenderness. UTS of cooked meat has been measured in relation to cooking temperature,[3,5,24,30] sarcomere length (including the effects of cold shortening)[5,11,37] and as a function of temperature below freezing on raw meat.[28] Tensile tests parallel and perpendicular to the muscle fibres have been taken to measure myofibrillar strength and connective tissue strength respectively.[5,30] However, there appears to have been little interpretation of the anisotropy of UTS in cooked muscle in relation to its qualitative fracture behaviour, so as to provide a quantitative explanation of why it breaks in the way it does. Voisey[41] pointed out that in much of the published work on tensile testing of cooked meat little attention has been given to the control of specimen size and strain rates used. Because of the viscoelastic nature of cooked meat, these factors, especially strain rate, would be expected to influence UTS results.[12]

Although UTS in itself is a fundamental property of any material, it need not necessarily provide a complete picture of fracture behaviour. For example, ordinary glass has a UTS of ≈ 170 MN m^{-2} and wood, across the grain, has a lower strength of ≈ 100 MN m^{-2},[13] but nevertheless, of the two, it is glass that we expect to be able to break more easily. In this case, UTS by itself is misleading. It is therefore necessary to extend our understanding of the fracture process using the concepts and approach of modern fracture mechanics.[21] Fracture mechanics determines toughness (toughness here being strictly defined in the engineering sense as the resistance of a material to the propagation of a crack or tear through it) in terms of the minimum energy required to propagate the fracture, or the concentration of stress at the tip of a crack or failure necessary for its propagation.[15] The energy criterion has proved of value when considering the fracture of tissues such as skin and artery wall.[31,32] A fracture mechanics approach is therefore a necessary and logical development in the study of the fracture behaviour of cooked meat.

4 MEAT FRACTURE—A DETAILED STUDY

To show how the study of meat fracture benefits from an approach based on relating basic and fundamental fracture measurements to structural aspects of fracture behaviour, a recent detailed study[33] will be described. This work was developed along the following lines:

a *Qualitative observations*: which part of the material structure breaks first, and where? It is important to appreciate which components of the material's structure are involved in fracture as this can suggest possible toughening mechanisms. Qualitative observations are therefore helpful in deciding which quantitative fracture measurements will be most informative, and are prerequisites of any attempts to model fracture behaviour.

b *Quantitative measurements*: (i) Ultimate tensile strength (UTS) is perhaps one of the most basic fracture properties. UTS can be measured in different orientations and qualitative aspects of fracture interpreted on this basis. (ii) The fracture toughness, in energetic terms, and notch sensitivity measurements are used to establish how structural features and resulting anisotropy in UTS determine the fracture behaviour of the material.

A Experimental material and specimen orientations

Figure 3 shows the form of specimens used in this work; a roughly cylindrical beef muscle (*M. semitendinosus*) was cut into thin slices (approximately 5 mm

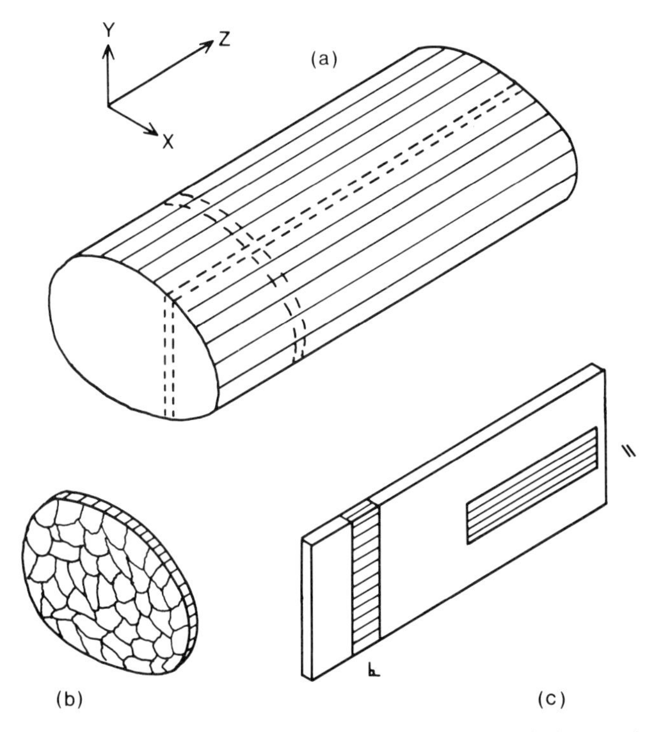

Fig. 3 Diagram to show (a) the pattern of cutting the whole muscle into (b) transverse and (c) longitudinal slices. Strips parallel (\\) or perpendicular (⊾) to the fibre direction are cut from the longitudinal slices (from Purslow, 1985)

thick) either transverse to the muscle fibre direction or parallel to the fibres running longitudinally along the muscle. All slices were cooked for 1 h at 80 °C. From the longitudinal slices, strips can be cut in two orientations; with the long axis of the strips parallel to the muscle fibre direction ("parallel" strips) and with the long axis of the strips perpendicular to the fibre direction ("perpendicular" strips), as shown in Fig. 3c.

B Qualitative observations

Detailed observations on the way this cooked meat breaks were made by carrying out tensile tests on notched strips in a variety of orientations and taking video recordings and still photographs of the fracture process. One of the most instructive test configurations to consider is propagation of fracture across a transverse strip, as shown in Fig. 4. A notch has been cut into the left-hand side, has opened up considerably and is growing from the right to the

Fig. 4 Propagation of rupture across a transverse specimen under tension. Structures (*S*) are strands of connective tissue (from Purslow, 1985)

left. Small scale ruptures or cavitations are occurring at muscle fibre boundaries, i.e. at the perimysium. These cavities are not confined to an area close to the moving crack tip, but are generalized. The macroscopic failure joins up a series of these small cavities, so that the path of fracture mainly lies around muscle fibre bundles. Bridging the crack tip there are strands (marked S on Fig. 4) of highly extensible connective tissue, which are the last structures to break. The observation is consistent with the view that the original small scale ruptures occurred at interfaces within, or at the surface of, the perimysium, leaving debonded planes of the connective tissue intact to form these bridging strands.

Closer examination of the fracture surface from such a test confirmed the impression of a strong perimysial involvement in fracture. Figure 5 is a low power micrograph of a fracture surface from a test in the same orientation as Fig. 4, which has been stained for collagen. The crack path has deviated through 90° in the centre of the path to run around muscle fibre bundle boundaries. Stained connective tissue is seen in both fracture surfaces produced from one tensile strip, showing that initial debonding or cavitation occurs within the connective tissue. The twisted and stained structure (S) is one of the bridging connective tissue strands that has finally broken and retracted back. These results indicate that primary failure occurred at the

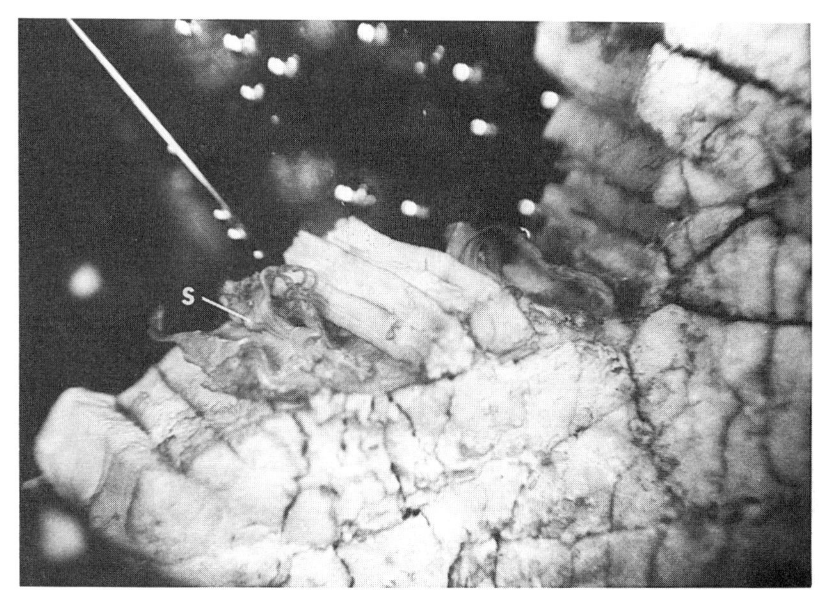

Fig. 5 Fracture surface of transverse specimen stained with aniline blue to show connective tissue. Structure (*S*) is broken and retracted connective tissue strand (from Purslow, 1985)

perimysial level only. This is supported by fracture observations in other test orientations. Figure 6 shows a longitudinal strip tested perpendicular to its muscle fibre direction; the separation of individual fibre bundles can be seen, with a network of fine connective tissue strands (C) bridging the fracture. Figure 7 shows a notched strip being pulled parallel to the fibre direction. The initial effect is merely to split back the perimysial interface perpendicular to the axis of the pre-cut notch. Eventually the debonded fibre bundles snap across.

Qualitatively, it seems that the perimysial connective tissue is an important substructure which to some extent determines the fracture behaviour of cooked meat. There is no great involvement of individual muscle fibres and their connective tissue, the endomysium; the debonding phenomenon occurs only at the perimysium under the conditions studied. These observations are in agreement with some SEM and video observations reported by Carroll *et al.*[8]

C Quantitative aspects

Ultimate tensile strength. This is simply defined as the maximum stress observed when a strip is loaded in tension until it breaks. In the light of

Fig. 6 Perpendicular strip of semitendinosus under tension, showing onset of separation of muscle fibre bundles with fine strands of connective tissue (C) bridging the gap (from Purslow, 1985)

qualitative observations on the fracture process, the most informative strength tests would be those that show the anisotropy of UTS in the longitudinal plane, i.e. to measure UTS parallel to the fibres and perpendicular to them. When testing perpendicular to the fibres no fibre bundles are broken; the material merely separates at the perimysium. On the other hand, when pulling strips parallel to the fibres, all the fibres in the cross-section must be broken.

Figure 8 shows the mean strength of perpendicular strips at three different strain rates. There is apparently a tendency for tensile strength to increase with strain rate, as may be expected from viscoelastic considerations,[12] but this trend is not significant ($P > 0.1$). Mean values of UTS lie in the region of 20–30 kN m^{-2}. Figure 9 shows the mean strength of strips tested parallel to the fibre direction at the same three strain rates. There is no effect of strain rate on tensile strength observable above the specimen to specimen variability which was considerable, despite the fact that all parallel and perpendicular strips come from the same muscle and were cut to the same size by template. The mean parallel strength on the range of strain rates tested is in the range of 300–400 kN m^{-2}. This is an order of magnitude above mean perpendicular UTS values, showing that there is considerable anisotropy in

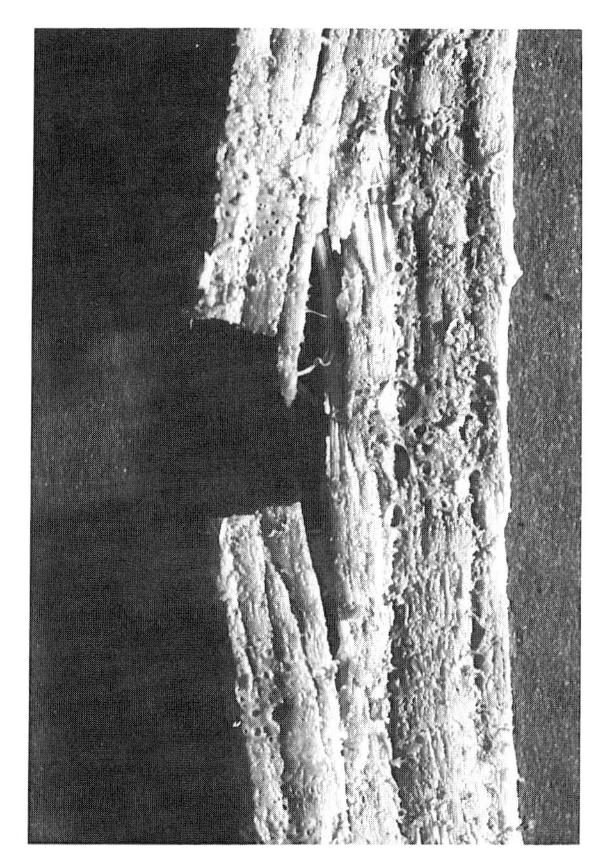

Fig. 7 Fracture across the fibre direction, showing the initial splitting back of perimysium (from Purslow, 1985)

strength between these two test directions; i.e. cooked meat is approximately ten times stronger when pulled along the fibre direction than perpendicular to it.

These strength results clearly show why cooked meat is observed to break in the way it does; the perimysium is much less strong than the bundles it surrounds in cooked meat.

The measured ratio of 10:1 or so of strength in the two directions is an interesting result in itself. Cook and Gordon[9] modelled the fracture behaviour of a composite material in which there were distinct, weak interfaces. By calculation of the pattern of stresses around an advancing crack tip (Fig. 10) they found that, if the strength across the interface, parallel to the crack axis was one fifth, or less, of the tensile strength of the surrounding

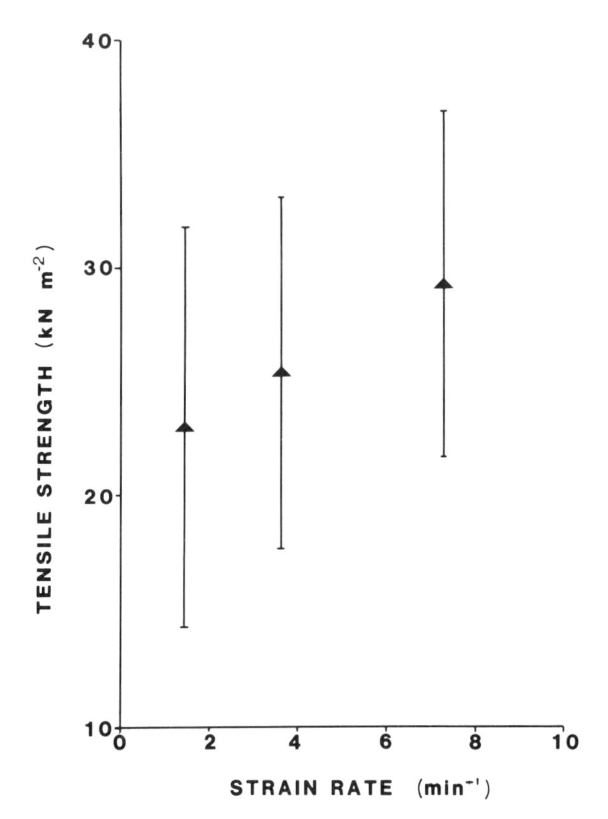

Fig. 8 Mean ultimate tensile strength perpendicular to fibre direction *v.* strain rate. Bars denote ± one standard error (from Purslow, 1985)

material perpendicular to the crack axis, then the interface would cavitate ahead of the crack tip. The crack would then run into this cavity and so blunt itself, rendering further crack propagation more difficult. This toughening mechanism, the Cook-Gordon mechanism, can be seen to operate in cooked meat by reconsidering Fig. 7. The measurement of anisotropy in tensile strengths parallel and perpendicular to the fibre direction shows why this mechanism occurs; the ratio of 10:1 or so is more than enough to satisfy Cook and Gordon's 5:1 criterion. Here again, quantitative measurements fit in with quantitative observations well.

D A fracture mechanics approach

Griffith[15] outlined two conditions necessary for a crack to grow in a stressed body and hence two criteria by which fracture toughness could be measured.

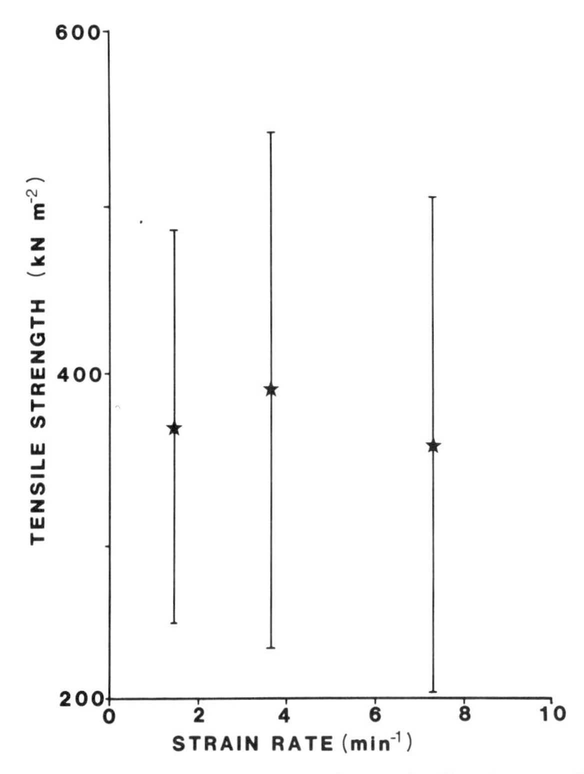

Fig. 9 Mean ultimate tensile strength *v.* strain rate in direction parallel to muscle fibres. Bars denote ± one standard error (from Purslow, 1985)

The first of these is the balance of energy consumed and released as a crack grows, and the second is the concentration of stress at the tip of the notch required to break local bonds. The stress concentrating effect of a crack had been shown previously by Inglis,[19] but it was Griffith's linkage of this effect with the thermodynamic consideration of energy exchanges that produced an explanation of why cracks could be stable under given loads and in doing so provided the basis for linear elastic fracture mechanics. Taking the energetic criterion first, consider a sheet of stressed material with a single edge-notch as shown in Fig. 11. Because the material is stretched, there is a certain amount of strain energy stored in the bulk of the sheet, except in the dotted area around the crack. The material in this zone is relaxed as the edges of the crack are free boundaries across which tension cannot be carried. If the crack grows by a small amount then new surfaces are produced at the crack tip, a process which consumes energy. However, the relaxed zone also grows as the length of the free boundaries (the edges of the crack) grows, a process which releases

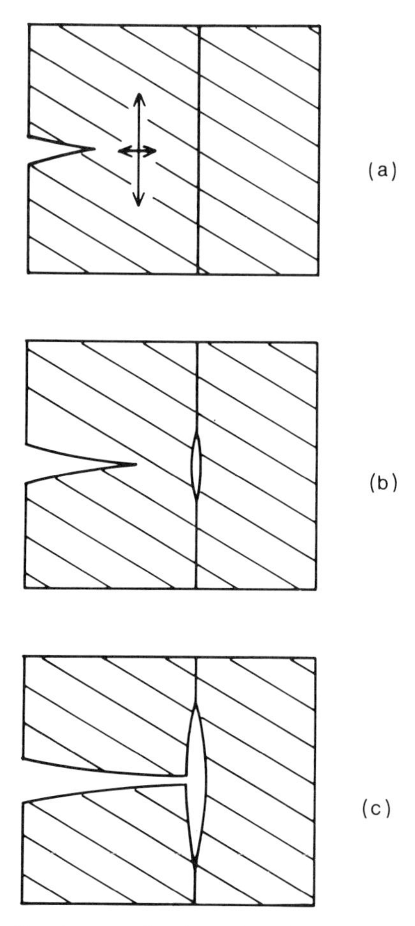

Fig. 10 The Cook Gordon mechanism. (a) A growing crack approaches an interface in the material. The arrows denote the relative intensity of the stress concentration effect parallel and perpendicular to the crack path. (b) The weak interface splits due to the stress component parallel to the crack path. (c) The advancing crack runs into this cavity and is blunted (redrawn from Gordon, 1980)

stored strain energy. If the rate of energy release with growing crack length from the growing relaxed zone is just enough to satisfy the rate of energy consumption with crack length by the creation of new surfaces, then the crack will propagate as it is energetically favourable for it to do so. If more than enough energy is released, the crack will accelerate and the excess energy will be used in heat, sound and in propelling parts of the specimen away from the fracture zone. This maximum energy release rate to feed the growing crack is

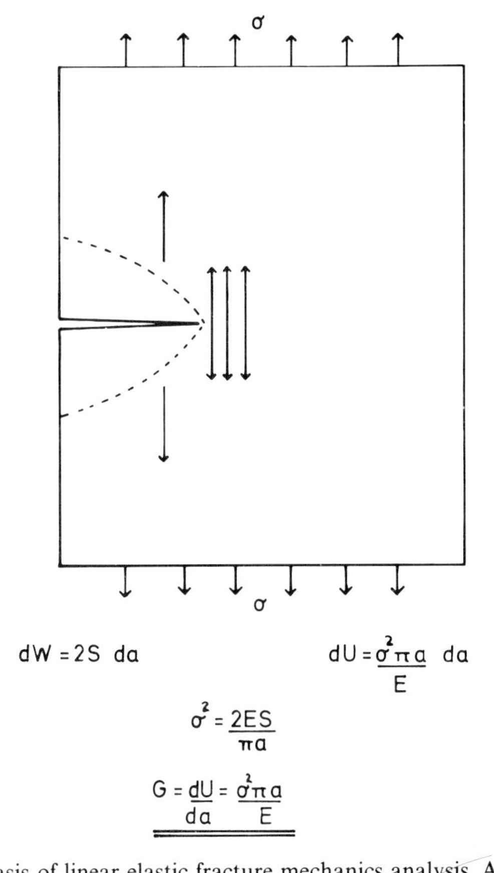

Fig. 11 The basis of linear elastic fracture mechanics analysis. A sheet of material with a single edge notch is shown with a remote applied stress σ. Arrows within the specimen boundaries represent force trajectories, which are more concentrated around the crack tip. There are no force trajectories and no stored energy within the dotted lines (relaxed zone). As crack length, a grows there is an expenditure of energy W on creating surfaces with surface energy S ($\mathrm{d}w = 2S\,.\,\mathrm{d}a$). There is also a change in stored strain energy U, given by $\mathrm{d}U = \dfrac{\sigma^2\pi a\,.\,\mathrm{d}a}{E}$. When these two expressions are equal, propagation will proceed; this will occur when $\sigma^2 = \dfrac{2ES}{\pi a}$, i.e. $\dfrac{\mathrm{d}U}{\mathrm{d}a} \geqslant 2S$. In non-brittle cases the surface energy is replaced by the more general term G

ideally a constant, characteristic of a particular material, the critical strain energy release rate, $\left(\frac{\partial U}{\partial a}\right)_c$ or G_c, where U is the total strain energy in the body, a is the crack length and the subscript c denotes the critical case. To measure this energetic toughness value we can make use of the finding that, in particular test piece configurations, the second derivative, $\frac{\partial^2 U}{\partial a^2}$ is negative.[16] In this condition the fracture propagation is very slow and controlled (quasi-static) and can be stopped at any point by unloading the specimen. The strain energy release rate is then close to the critical value at all points in the propagation. As the total work, U, expended in propagating a crack from initial length a_1 to a new length a_2 through a sheet of material of thickness t is given by:

$$U = \int_{a_1}^{a_2} G.t.\mathrm{d}a$$

and $G \simeq G_c$ at all stages

$$U = G_c.t(a_2 - a_1)$$

$t.(a_2 - a_1)$ is the area through which the crack has been propagated, A_p, and so $G_c = \frac{U}{A_p}$ i.e., G_c may be evaluated as the total work done per unit area of material cleaved by the passage of the crack. Rivlin and Thomas[34] outlined two such quasi-static propagation methods for nonlinear, finite strain materials, specifically rubbers. These methods have been adapted to measure work of fracture of animal tissues.[31–33]

The work of fracture for cracks propagating parallel to the muscle fibre direction in longitudinal strips ("along fibres") and for cracks propagating around fibre bundles across a transverse slice have been determined for cooked beef *M. semitendinosus*,[33] and mean values are presented in Table 1. Both values are for propagation of cracks through perimysial material and are relatively low in absolute terms, explaining why the perimysium is the preferred fracture site. The difference between the two values may be ascribed to three factors. Firstly, the more tortuous crack path around muscle fibre bundles across the transverse slice results in a greater actual area cleaved per nominal cross-sectional area. Secondly, cavitation at the fibre bundle boundaries only occurs in the transverse orientation and is a process which irreversibly absorbs energy, and thirdly, differences in loading configurations between the two tests may result in different amounts of generalized plasticity in the bulk of each specimen. The work of fracture for cracks running across and through the muscle fibres would appear to be far greater than the values

Table 1 Works of fracture along and between fibres (from Purslow, 1985)

	Along fibres	Between fibres
Mean work of fracture/kJ m^{-2}	0.42	1.83
Standard error	0.035	0.13
Number of specimens	19	17

reported for perimysial fracture in Table 1, as any attempt to propagate cracks in this direction results in the crack deviating to run along the weaker fibre bundle boundaries, at right angles to the desired path.

E Notch sensitivity

Griffith[15] identified a second condition necessary for a crack to propagate; that there should be sufficient stress intensity around the tip of the crack to break atomic bonds by the local concentration of sufficient strain energy. The crack can be viewed as a mechanism for strain energy transfer from the bulk of the specimen to the fracture site at the tip of the crack. This transfer efficiency or stress concentration effect depends on the shape of the crack. For a linearly elastic material:[20]

$$\sigma_L = \sigma(1 + 2a/\rho)$$

where σ_L is the maximum local stress at the crack tip, σ the remote applied stress, a the crack length and ρ the crack tip radius of curvature.[20,21] Thus, for a given crack length, a sharp crack tip (i.e. a small radius) results in a high local stress concentration, which effectively means good strain energy transfer. If, however, the crack tip is massively blunted then strain energy transfer will be much poorer and so propagation is much more difficult. Such crack blunting occurs when trying to propagate cracks across the muscle fibre direction in cooked meat due to the Cook-Gordon mechanism, as previously seen. The effect of this is outlined in Fig. 12.[14]

In Fig. 12*A* there is a unaxial array of elements that are very well bonded together with strong interfaces. Strain energy released by the fracture of an element is able to be transferred by shear across the interfaces to the crack tip, so as to influence the fracture of successive elements. This is referred to as notch-sensitive behaviour.[20] Griffith's formula relates nominal stress at fracture to crack length for a brittle continuum in plane stress:

$$\sigma_B = \left(\frac{G_c . E}{\pi a}\right)^{1/2}$$

where σ_B is the nominal breaking stress, G_c the critical strain energy release

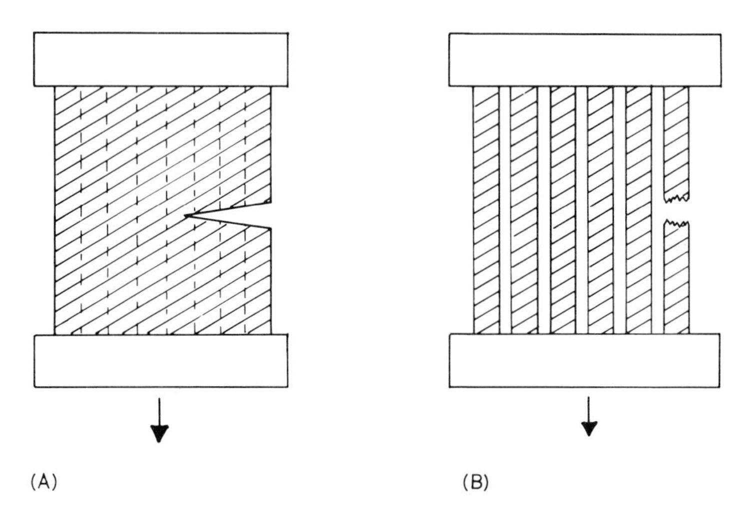

Fig. 12 Crack propagation in continuous and sub-divided structures. (A) Elements joined with well-bounded interfaces; crack propagates. (B) Elements isolated, crack is stopped (redrawn from Gordon, 1980)

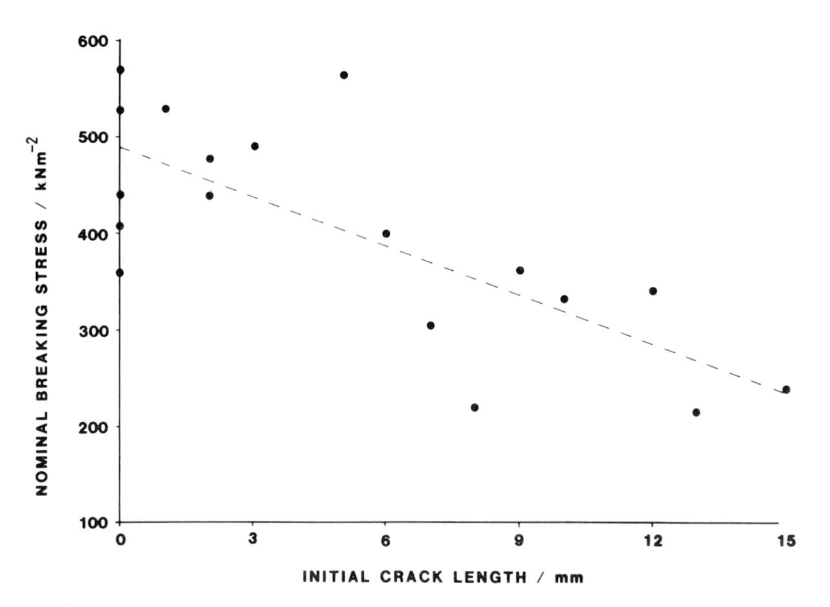

Fig. 13 Notch sensitivity of cooked beef *M. semitendinosus*. Nominal breaking stress is shown as a function of the initial length of notch cut across the fibres (from Purslow, 1985)

rate, E the Young's modulus and a the crack length. From this it is apparent that for completely notch sensitive behaviour, i.e. brittle fracture in a continuum, σ_B is proportional to $(1/a)^{1/2}$. Hedgepeth[18] calculated a notch sensitivity relationship for uniaxial filamentous arrays which is of a similar general shape to the continuum case.

On the other hand, in Fig. 12*B*, if there is very little or no interfacial integrity then there is no mechanism for the transference of energy from one broken element to effect the fracture of the next and the elements fail independently. The assemblage of elements is completely insensitive to the presence of a notch and the nominal breaking stress decreases linearly, in simple proportion to the remaining, intact cross-sectional area:[20]

$$\sigma_B = \sigma_N(1 - a/W)$$

where σ_N is the nominal breaking strength of the material and W the total width.

Figure 13 shows the relationship between nominal breaking stress and crack length in cooked beef strips, with the crack direction perpendicular to the fibre direction. It is apparent that this material shows notch-insensitive behaviour. There appears to be little or no shear communication of strain energy from fibre bundle to bundle, showing why it is difficult to propagate fracture across the fibre direction.

5 CONCLUSIONS

A fracture mechanics approach to the fracture behaviour of cooked meat has provided quantitative evaluations which have concurred with qualitative observations in a manner which elucidates some of the mechanisms responsible for the toughness of this material.

The perimysium appears to be an important level of structural organization in cooked beef muscle in relation to its fracture behaviour. The low transverse strength and work of fracture of this connective tissue layer results in the overall structure being easily separable into individual fibre bundles, but propagation of fracture across the fibre direction is much more difficult. Limitations in the applicability of the classical brittle Griffith approach are demonstrated by the notch-insensitivity observations. The qualitative aspects of the fracture behaviour of this material are accounted for by a model of cooked meat as a uniaxial fibrous composite of strong fibre bundles in a weak connective tissue "matrix" with poor interfacial strength between the two components.

References

1. Bailey, A. J. and Sims, T. J. (1977). *J. Sci. Fd. Agric.* **28**, 565.
2. Boccard, R., Buchter, L., Casteels, E., Cosentino, E., Dransfield, E., Hood, D. E., Joseph, R. L., Macdougall, D. B., Rhodes, D. N., Schon, I., Tinbergen, B. J. and Touraille, C. (1981). *Livestock Production Sci.* **8**, 385.
3. Bouton, P. E. and Harris, P. V. (1972). *J. Fd Sci.* **37**, 140.
4. Bouton, P. E. and Harris, P. V. (1972). *J. Fd Sci.* **37**, 218.
5. Bouton, P. E., Harris, P. V. and Shorthose, W. R. (1975). *J. Texture Studies* **6**, 297.
6. Bratzler, L. J. (1932). Measuring the tenderness of meat by means of a mechanical shear. MSc Thesis, Kansas State College.
7. Bratzler, L. J. (1949). *Proc. Ann. Reciprocal Meat Conf.* 117.
8. Carroll, R. J., Rorer, F. P., Jones, S. B. and Cavanaugh, J. R. (1978). *J. Fd. Sci.* **43**, 1181.
9. Cook, J. and Gordon, J. E. (1964). *Proc. Roy. Soc. Lond.* **A282**, 508.
10. Davey, C. L. and Gilbert, K. V. (1969). *J. Fd. Sci.* **34**, 69.
11. Davey, C. L. and Gilbert, K. V. (1977). *Meat Sci.* **1**, 49.
12. Ferry, J. D. (1980). Viscoelastic properties of polymers, 3rd edn, John Wiley, New York.
13. Gordon, J. E. (1978). Structures. Penguin, Harmonsworth, UK.
14. Gordon, J. E. (1980). In The Mechanical Properties of Biological Materials (34th Symposium of the Society for Experimental Biology). ed. J. F. V. Vincent and J. D. Currey, p. 1–11, Cambridge University Press.
15. Griffith, A. A. (1921). *Phil. Trans. Roy. Soc. Lond.* **A221**, 163.
16. Gurney, C. and Hunt, J. (1967). *Proc. Roy. Soc. Lond.* **A299**, 508.
17. Hammond, J. J. (1940). *J. Soc. Chem. Ind.* **59**, 521.
18. Hedgepeth, J. M. (1961). Stress concentrations in filamentary structures. NASA TND-882.
19. Inglis, C. E. (1913). *Proc. Inst. Naval Architects. Lond.* **60**, 219.
20. Kelly, A. (1966). Strong Solids, Clarendon Press, Oxford.
21. Knott, J. F. (1973). Fundamentals of Fracture Mechanics, Butterworths, London.
22. Kramer, A. (1961). *Fd Scientst* **5**, 7.
23. Kramer, A., Burkhardt, G. J. and Rogers, H. P. (1951). *Canner* **112**, 34.
24. Locker, R. H. and Carse, W. A. (1976). *J. Sci. Fd. Agric.* **27**, 891.
25. Locker, R. H. and Wild, D. J. C. (1982). *J. Texture Studies* **13**, 71.
26. Macfarlane, P. G. and Marer, J. M. (1966). *Fd Technol.* **20**, 134.
27. Marsh, B. B. and Leet, N. G. (1966). *J. Fd Sci.* **31**, 450.
28. Munro, P. A. (1983). *Meat Sci.* **9**, 43.
29. Poole, M. F. and Klose, A. A. (1969). *J. Fd. Sci.*, **34**, 524.
30. Penfield, M. P., Barker, C. L. and Meyer, B. H. (1976). *J. Texture Studies* **7**, 77.
31. Purslow, P. P. (1983). *J. Mater. Sci.* **18**, 3591.
32. Purslow, P. P. (1983). *J. Biomechanics* **16**, 947.
33. Purslow, P. P. (1985). *Meat Sci.* **12**, 39.
34. Rivlin, R. S. and Thomas, A. G. (1953). *J. Polymer Sci.* **10**, 291.
35. Rowe, R. W. D. (1981). *Tissue and Cell* **13**, 681.
36. Sale, A. J. H. (1960). In Texture in Foods. SCI monograph No. 7, pp. 103–108. SCI, London.
37. Stanley, D. W. and Swatland, H. J. (1976). *J. Texture Studies* **7**, 65.
38. Stanley, D. W., Pearson, G. P. and Coxworth, V. E. (1971). *J. Fd Sci.* **36**, 256.
39. Stanley, D. W., McKnight, L. M., Hines, W. G. S., Usborne, W. R. and DeMan, J. M. (1972). *J. Texture Studies* **3**, 51.

40. Szczesniak, A. S. and Torgeson, K. W. (1965). *Adv. Food Res.* **14**, 33.
41. Voisey, P. W. (1976). *J. Texture Studies* **7**, 11.
42. Voisey, P. W. and Larmond, E. (1974). *J. Can. Inst. Food Sci. Technol.* **7**, 243.
43. Volodkevich, N. N. (1983). *Fd Res.*, **3**, 221.
44. Warner, K. F. (1928). *Proc. Am. Soc. Animal Production* **21**, 114.
45. Winkler, C. A. (1939). *Can. J. Res.* **17D**, 8.

10 Modelling of Heat and Mass Transfer in Foodstuffs

P. B. JOHNS and S. H. PULKO

Department of Electrical and Electronic Engineering, University of Nottingham

1 INTRODUCTION

Much of the processing of foodstuffs such as freezing, cooking or drying involves transfer of heat or mass within the foodstuff material. The mechanism of transfer of heat is essentially the same as that of mass and is governed predominantly by the two basic laws of diffusion.

The first of these basic laws says that the heat flux density (mass flow density) J across the element shown in Fig. 1, is related to the gradient of the temperature (pressure) across it. In one space dimension this is

$$J = -K\frac{\partial V}{\partial x} \tag{1}$$

where J is in J m^{-2} s^{-1}, V is in deg K and K is the thermal conductivity which is in W K^{-1} m^{-1} and may be a constant or vary with x and/or V.

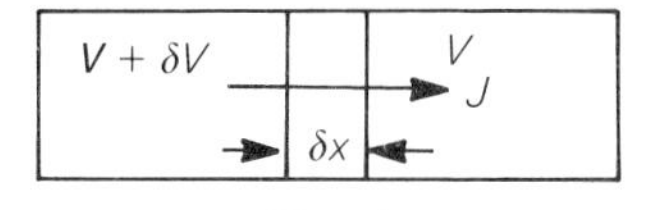

Fig. 1

FOOD STRUCTURE AND BEHAVIOUR
ISBN 0-12-104230-8

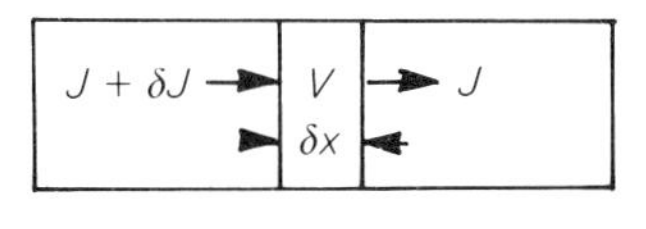

Fig. 2

The second of the basic laws accounts for the difference in flux or flow across the element shown in Fig. 2 and says that accumulation of heat (mass) corresponds to an increase in temperature (pressure) V with time. Thus, in one dimension

$$\frac{\partial V}{\partial t} = \frac{1}{S}\frac{\partial J}{\partial x} \tag{2}$$

where for heat transfer S is the specific heat capacity in J K^{-1} m^{-3} and for mass transfer S and V are usually combined to give a concentration parameter instead of V.

Combining equations 1 and 2 for constant K gives the well known diffusion equation

$$\frac{\partial V}{\partial t} = \frac{K}{S}\frac{\partial^2 V}{\partial x^2} \tag{3}$$

or more generally for two and three space dimensions

$$\frac{\partial V}{\partial t} = \frac{K}{S}\nabla^2 V \tag{4}$$

In most practical situations problems can be nonlinear, inhomogeneous or have complicated boundaries in two and three dimensions and usually it is not possible to obtain an analytical solution to the diffusion equation. It is then necessary to adopt a modelling procedure to allow the problem to be simulated on a computer, and this paper shows how simple and efficient models can be generated.

2 LUMPED NETWORK MODELLING

Consider the cube of foodstuff material shown in Fig. 3 and assume that the thermal conductivity throughout is K and the specific thermal capacity is S. The three-dimensional network of resistors R and the capacitor C can represent the cube if

$$2R = \frac{1}{K}\frac{1}{\Delta l} \tag{5}$$

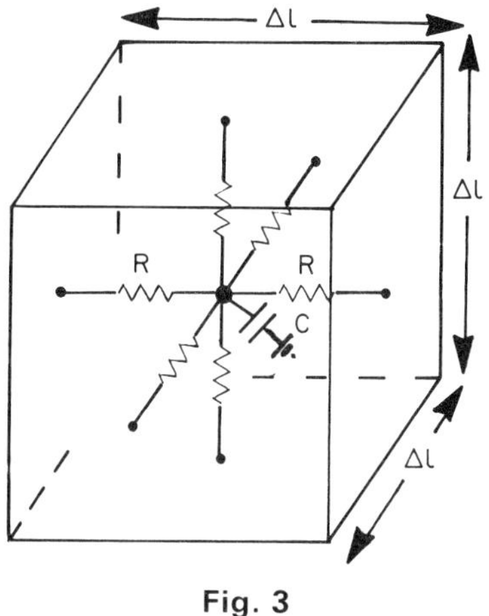

Fig. 3

and

$$C = S\,\Delta l^3 \tag{6}$$

The heat flux density travelling through the cube in a given direction can now be represented by a heat flow, I, through the resistors where

$$I = J\,\Delta l^2 \tag{7}$$

By placing blocks of material together with suitable boundary terminations it can be seen that the material shape and its thermal properties can be modelled by a lumped network. Solution of the lumped network then provides an approximate solution of the physical problem. For example, the network shown in Fig. 4 models one-dimensional diffusion and the network

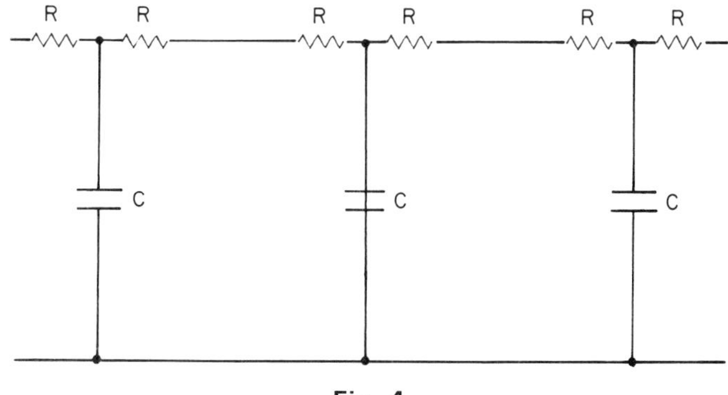

Fig. 4

equations may be written as follows,

$$I_n = \frac{1}{2R}[V_{n-1} - V_n] \tag{8}$$

$$I_n - I_{n+1} = C\frac{\partial V_n}{\partial t} \tag{9}$$

Clearly a set of ordinary differential equations can be formed for the network as a whole, and a typical one is:

$$\frac{\partial V_n}{\partial t} = \frac{1}{2RC}[V_{n-1} - 2V_n + V_{n+1}]. \tag{10}$$

Using equations 5 and 6 we have:

$$\frac{\partial V_n}{\partial t} = \frac{K}{S}\frac{1}{\Delta l^2}[V_{n-1} - 2V_n + V_{n+1}] \tag{11}$$

which is the diffusion equation (equation 3) with the spatial differential differenced.

Differencing of the time differential in equation 11 yields

$$_{k+1}V_n = {_kV_n} + \frac{\Delta t}{2RC}[_kV_{n-1} - 2_kV_n + {_kV_{n+1}}] \tag{12}$$

where $_kV_n$ is the temperature at node n at timestep k, and this is a simple finite difference routine for diffusion problems.

While the spatial discretization has physical meaning and lends itself to intuitive modelling, the temporal differencing has little physical meaning. Indeed if $\Delta t/2RC > 0.5$ then the routine becomes unstable and has no physical meaning at all. The process of transmission-line modelling renders both space and time discrete while keeping a physical meaning and so instability does not occur.

3 TRANSMISSION-LINE MODELLING (TLM)

Transmission-line modelling uses a transmission-line network of nodes connected by transmission-lines. These transmission-lines are loss-free tubes which carry heat or mass pulses a finite distance Δl in a finite time Δt. The transmission-lines have cross-sectional geometry such that a pulse of heat travelling along the tube has associated with it a proportional temperature or, in the mass transfer case, the pulse of mass has an associated pressure. Thus a transmission-line tube with given amount of mass travelling down it will have a low pressure if the diameter is large and vice-versa. The ratio of the temperature to the heat flow (or pressure to the mass flow) in the tube is termed the characteristic resistance of the transmission-line.

The resistors modelling the thermal conductivity are now clustered around the node as shown in Fig. 5, and the transmission-lines connect them to neighbouring nodes. The values of the resistors are the same as for the lumped network of course, but this time the thermal capacity is modelled by the capacity of the transmission-line tube. The basic network equation for a capacitor is

$$C = \frac{Q}{V}$$

Since current is given by $J = \dfrac{Q}{\Delta t}$ the impedance of the transmission line is related to C by

$$Z = \frac{V}{J} = \frac{\Delta t}{C}$$

In the case of heat flow

$$Z = \frac{V^i}{I^i} = \frac{\Delta t}{C} \tag{13}$$

where I^i represents the heat flux density and V^i the associated temperature.

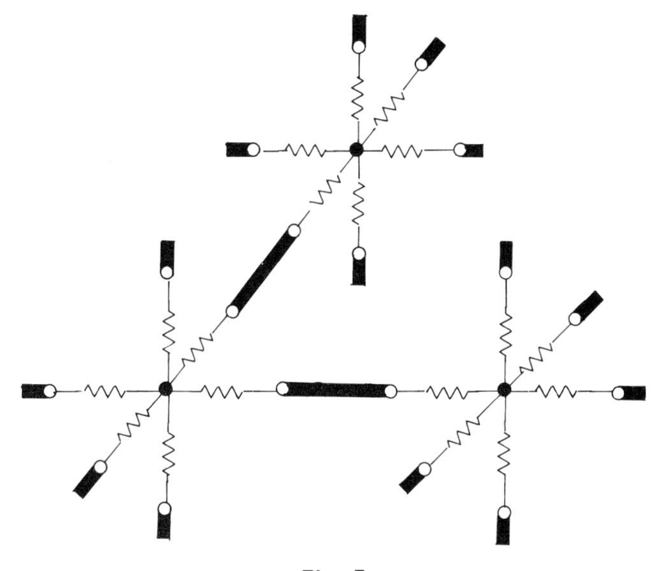

Fig. 5

Solution of the network takes place by considering heat flow pulses with the temperature $\mathbf{V}^i$ incident simultaneously at all of the terminals of all of the nodes. These are scattered instantaneously into reflected pulses $\mathbf{V}^r$ according to the scattering equation

$$ {}_k\mathbf{V}_n^r = \mathbf{S}_k\mathbf{V}_n^i \tag{14}$$

The pulses then travel along the transmission-lines to the resistor terminals of neighbouring nodes to become incident pulses on the neighbouring nodes.

$$ {}_{k+1}\mathbf{V}^i = \mathbf{C}_k\mathbf{V}^r \tag{15}$$

C is a connection matrix which describes which resistor terminal number is connected to which, through a transmission-line. Repeated operation of equations 14 and 15 forms the TLM algorithm and the computer follows the repeated scattering and travelling of pulses to neighbouring nodes. Suppose that mass transfer is taking place in the form of atoms on a crystal lattice, then clearly the TLM process has the physical meaning of atoms jumping from location to location.

The form of **C** is straightforward; it consists of entries of unity off the diagonal to carry pulses to neighbouring nodes, and on the diagonal to return pulses to the same node. The burning question is how to find **S**. The answer is by recognizing that the node must conserve heat energy (or mass) and the temperature drop across the node must be related to the heat flow through the resistors. Consider the single node in a one-dimensional network shown in Fig. 6, which also shows the incident pulses before scattering and the scattered pulses after scattering. The scattering equation is

$$\begin{bmatrix} V_1^r \\ V_2^r \end{bmatrix} = \begin{bmatrix} S_{11} & S_{12} \\ S_2 & S_{22} \end{bmatrix} \begin{bmatrix} V_1^i \\ V_2^i \end{bmatrix} \tag{16}$$

where S_{12} represents scattering into region 1 from region 2. Suppose that $V_2^i = 0$ then the total heat flow into terminal 1 is

$$\frac{V_1^i}{Z}[1 - S_{11}]$$

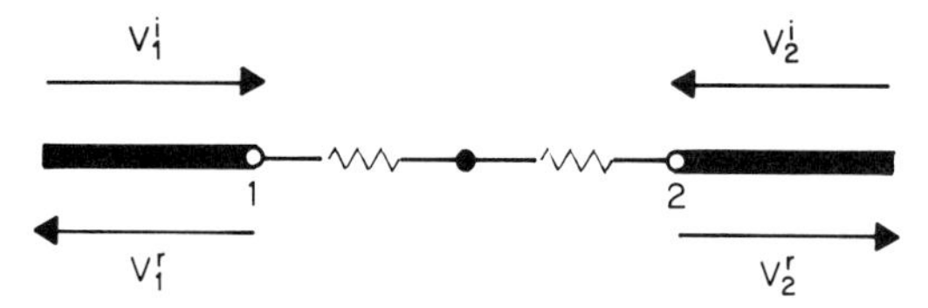

Fig. 6

Equating this to the heat flow leaving terminal 2

$$\frac{V_1^i}{Z}[1 - S_{11}] = \frac{V_1^i}{Z} S_{21}$$

i.e.

$$1 - S_{11} = S_{21} \tag{17}$$

The temperatures at terminals 1 and 2 are $V_1^i[1 + S_{11}]$ and $V_1^i S_{21}$ respectively and so the equation for the temperature drop is

$$\frac{V_1^i[1 + S_{11}] - V_1^i S_{21}}{V_1^i[1 - S_{11}]/Z} = 2R$$

i.e.

$$Z[1 + S_{11} - S_{21}] = 2R[1 - S_{11}] \tag{18}$$

The scattering matrix is symmetrical and so equations 17 and 18 give the solution:

$$\mathbf{S} = \frac{1}{R + Z}\begin{bmatrix} R & Z \\ Z & R \end{bmatrix} \tag{19}$$

and from equation 13

$$Z = \frac{\Delta t}{C} \tag{20}$$

The important point to note is that in TLM the transmission-line model that has been formed in the process of modelling is solved exactly in the computer. This is not true of the lumped network where the time differencing introduces an additional approximation. Further, since the transmission-line model conserves the flow quantity and contains no energy producing elements, the method must be stable for all values of Δt. The accuracy can decrease with increasing Δt but stability is not lost. Also it should be noted, that the method is a single step method in that the pulses at timestep $k + 1$ are calculated in terms of those at timestep k only. The accuracy, stability and consistency of TLM have been examined in some detail in references 1 and 2.

It is of some interest to look at equation 19 for the case where $Z = R$. The scattering process is now such that pulses incident on a node are scattered half into the reverse direction and half into the forward direction, and it can be shown that the TLM routine is then identical to equation 12 with $\Delta t/2RC = 1/2$. For other values of Z (other values of Δt) and for the simple linear and homogeneous example here the TLM method is the same as the two-step method of Du Fort and Frankel.[2] In general, however, TLM methods cannot be realized as finite difference routines.

It is often convenient to place boundaries in the TLM network half way between the nodes. Figure 7 shows a general boundary termination where the

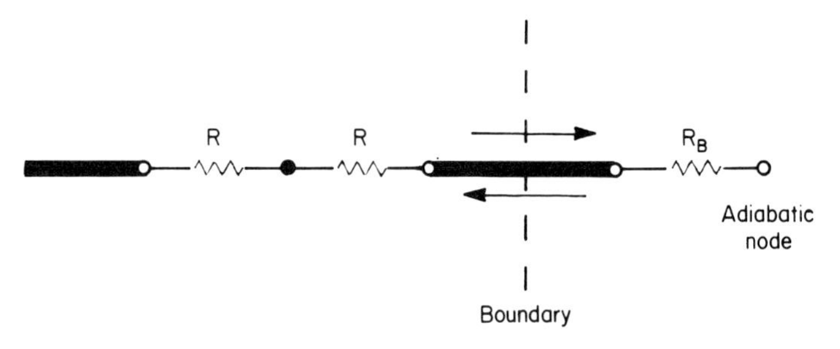

Fig. 7

transmission line is connected to a fixed temperature node at temperature V_B through a boundary resistance R_B. The pulse V^i is incident and the problem is to find the reflected pulse V^r. The product of the resistance R_B and the heat flow through it must equal the temperature drop across it and so:

$$\frac{R_B}{Z}(V^i - V^r) = V^i + V^r - V_B$$

thus

$$V^r = \frac{(R_B - Z)}{(R_B + Z)} V^i + \frac{Z}{(R_B + Z)} V_B \tag{21}$$

If the boundary is insulating then R_B is infinite and

$$V^r = V^i$$

If the boundary is an infinite sink then $R_B = 0$ and

$$V^r = -V^i + V_B$$

4 EXAMPLES OF TLM OF HEAT TRANSFER

A One dimensional Cartesian diffusion

1 AN INFINITE ROD

Probably the simplest modelling situation is that of a uniform rod whose thermal properties do not vary with temperature, perfectly insulated along its length, in the centre of which a pulse of heat is injected. In this case, if the rod is sufficiently long and the timescale over which modelling is attempted is sufficiently short, then all boundary and end effects may be ignored and the rod assumed to be infinitely long.

Figure 8 illustrates the mechanism of TLM for the first three timesteps in

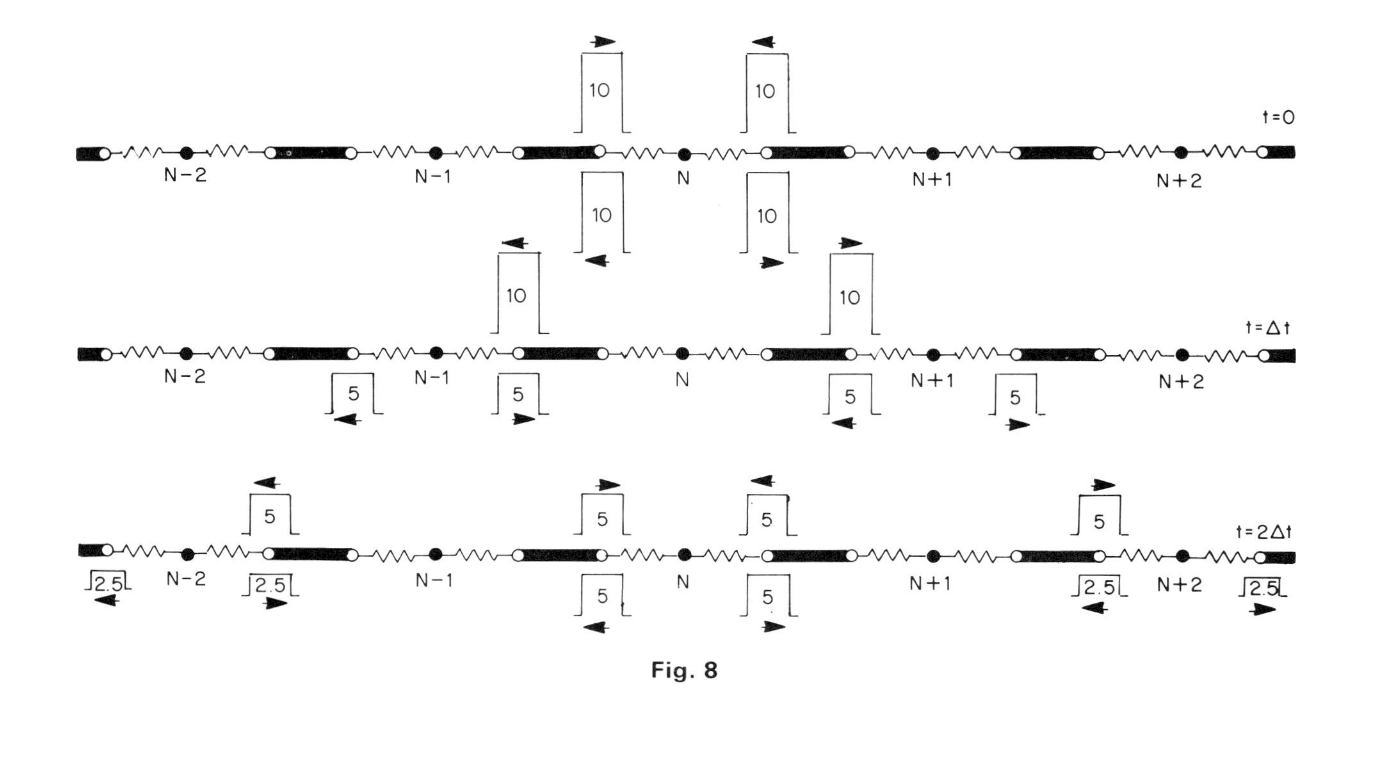

Fig. 8

this simple situation. At time $t = 0$ a pulse of temperature is injected at node N. Scattering occurs and, to take the simplest case of $R = Z$, half of each pulse is scattered in each of the forward and backward directions. This process gives rise to two reflected pulses, one at each terminal, of height 10 units. The temperature of node N is the mean of the temperatures of its two terminals. Since scattering is instantaneous the total pulse height at each terminal of node N is 20 units. This gives a value of 20 for the temperature of node N. During the time interval Δt the reflected pulses travel along the transmission lines and become incident pulses at nodes $(N + 1)$ and $(N - 1)$. In the absence of another injected pulse there is now no incident pulse at either terminal of node N. Nodes $(N + 1)$ and $(N - 1)$ both have a single incident pulse and the second scattering event gives rise to two reflected pulses at each of these nodes, for the case of $R = Z$, each of height 5 units. This process gives rise to a temperature distribution of 10 units at node $(N - 1)$, zero at node N, and 10 units at node $(N + 1)$. During the interval Δt one reflected pulse from each node travels towards the central node (node N), while the other travels to the nearest previously unconsidered node. By the third scattering event, therefore, the system has extended to include five nodes with a corresponding reduction in pulse heights. Thus, in heat flow terms, heat is being conducted away from the central injection point towards the extremities of the rod.

Figure 9 shows a plot of temperature versus distance from the injection point, computed using the TLM method for an infinitely long rod with $S = 500\,\mathrm{J\,kg^{-1}\,K^{-1}}$, $\rho = 1000\,\mathrm{kg\,m^{-3}}$ and $K = 20\,\mathrm{J\,s^{-1}\,K^{-1}}$. Δl was taken as 0.001 m and the timestep, Δt, was 0.001 s. The initial temperature pulses were of height 250 °C and the initial rod temperature 0 °C. It is clear from the foregoing discussion that the curve obtained by joining the maxima in Fig. 9 would correspond to the injection of twice the input of heat, i.e. another two pulses would be necessary to fill in the total area under the curve. Therefore, if the temperature values obtained are divided by two, the results should correspond to those found analytically by solution of the heat equation

$$\frac{\partial \theta}{\partial t} = D \frac{\partial^2 \theta}{\partial x^2} \qquad \text{where } D = \frac{K}{S\rho}$$

Figure 10 shows the two sets of results, the circles representing values obtained by TLM and the continuous line the analytical data. The agreement between the two methods is very good, the maximum error occurring, as would be expected, in the region of maximum rate of change. This error could be reduced by using a smaller value of Δt. Figure 11 illustrates the operation of a similar routine using three different timesteps, $\Delta t = 1.25$ s being the stability limit of the corresponding finite difference routine.

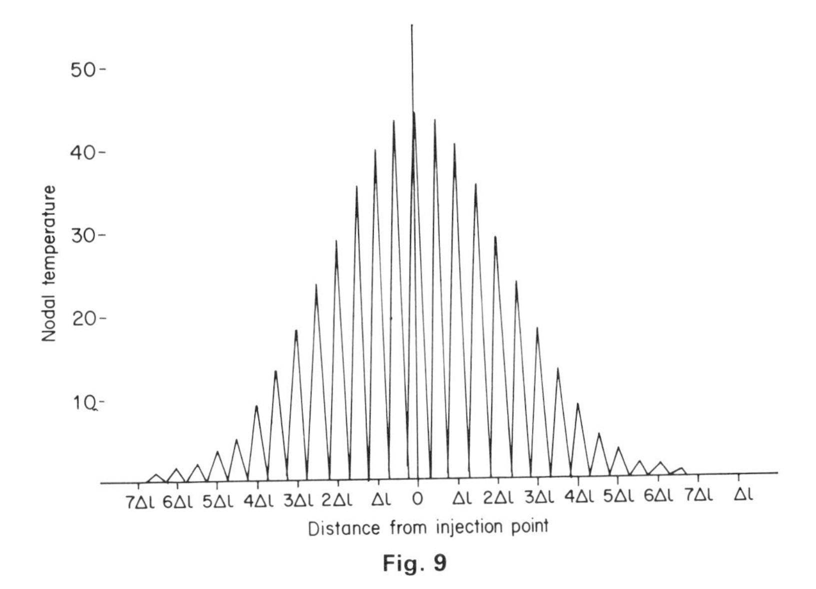

Fig. 9

An extension of this example is to consider the temperature distribution in a bar of finite length where heat is injected at each end from a gaseous atmosphere at constant temperature. In these circumstances the heat flux, q, from the gaseous atmosphere into one end of the bar is given by

$$q = -h(T_g - T_m)$$

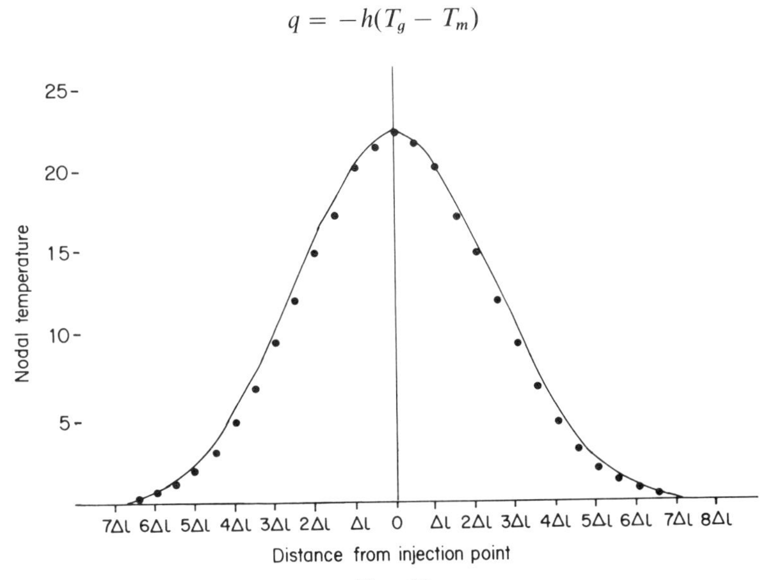

Fig. 10

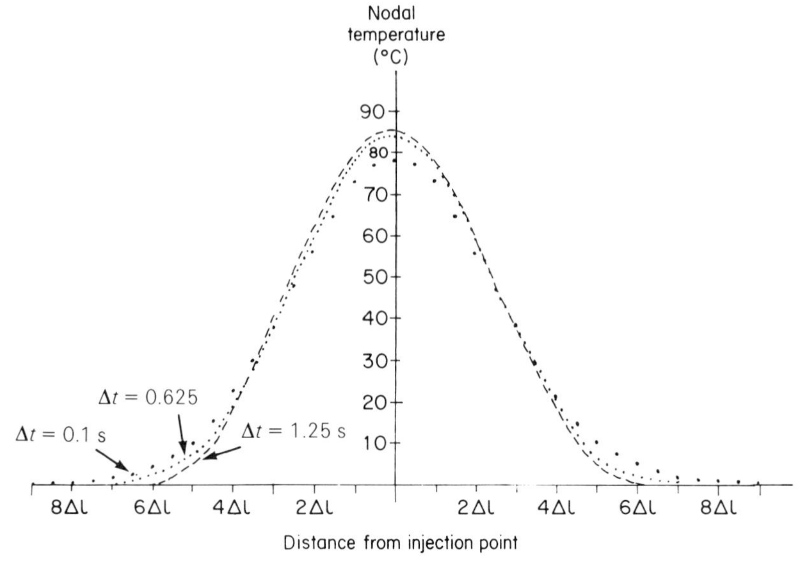

Fig. 11

where T_g and T_m are temperatures of the gas and material respectively and h is the heat transfer coefficient in W K^{-1} m^{-2}. Heat transfer between the environment and the bar may be represented by current flow through a boundary resistance. The value of the boundary resistance is given by:

$$R_B = \frac{1}{h(\Delta l)^2} \tag{22}$$

where $(\Delta l)^2$ is the cross sectional area of the bar. If the bar is considered to be made up of cubes of side Δl then the resistances and capacitances required to model heat flow in the bar are given by equations 5 and 6 and the scattering matrix by equation 19. Scattering at the boundaries is described by equation 20. Figure 12 represents the temperature distribution along a bar of length 8 cm with identical thermal parameters to the infinite bar considered earlier. In this case one end of the bar received input from an environment at 400 °C and the other from an environment at 200 °C. The heat transfer coefficient, h, was taken to be 400 W K^{-1} m^{-2} for both ends of the bar and nodes were located every 0.01 m. Ten nodes were used in all, two acting as adiabatic nodes situated in the atmosphere, and the boundaries were taken to be half way between nodes.

2 A SPHERE

Spherical symmetry renders the modelling of heat flow in a spherical body a

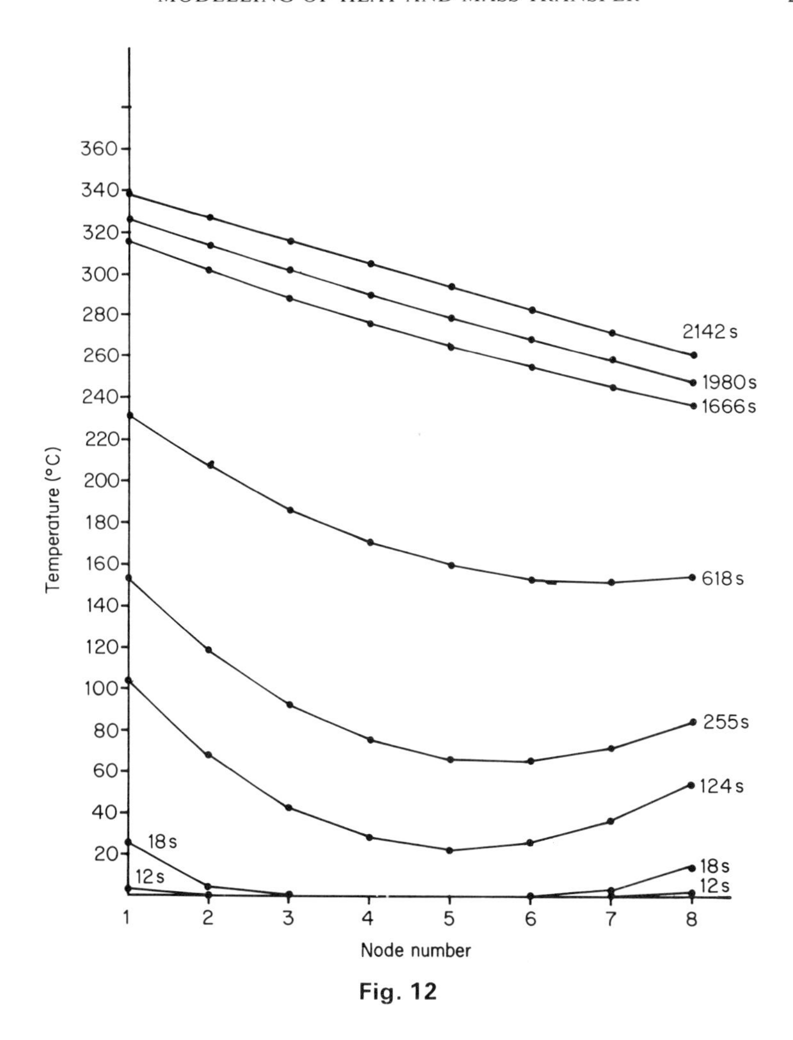

Fig. 12

one-dimensional problem. The sphere is considered to be composed of a series of shells surrounding a central core. This physical model can then be represented by a one-dimensional transmission line network.

Figure 13 shows a section through a sphere divided into six shells and a central core. The solid lines represent boundaries and each shell has a scattering node half way through its wall thickness. Clearly the appropriate values of R and C will be different for each node and they must be calculated individually. For nodes 1 to 6 the appropriate value of resistance is given by $\int \frac{\mathrm{d}r}{4\pi r^2}$ between the limits of the inner and outer radii of the shell. For node 4 this means that:

$$2R = -\left[\frac{1}{4\pi r}\right]_{r=B}^{r=A}$$

The resistance corresponding to node 7, the adiabatic node, takes account of heat flow from the atmosphere into the sphere given by

$$R_B = \frac{1}{h \times \text{surface area of sphere}}$$

The capacitance corresponding to nodes 2 to 6 involves the volume of a shell of thickness Δl. The region appropriate to node 4 is shaded in Fig. 13 and the capacitance is derived from:

$$C = \text{Volume of shell} \times S \times \rho$$

Figure 14 represents the terminations of the transmission line network used to model the sphere. The transmission line connected to node 7 represents the thermal capacity of the outer shell of thickness $\Delta l/2$, which is the outer shaded region in Fig. 13. The shaded central circle in Fig. 13 represents the core of the shell model and in the transmission line network its capacitance is modelled by a stub transmission line. A stub is an open circuit transmission line, which reflects pulses from the open circuit without change of phase. The pulse height at the open circuit end of the transmission line is, therefore, twice the height of the pulse incident on the stub from node 1. The capacitance of the stub is simply

$$C_{\text{stub}} = \text{Volume of core} \times S \times \rho$$

but since, in a period Δt, the pulse travels to the open circuit end of the stub and back to node 1, equation 13 is modified to:

$$Z_{\text{stub}} = \frac{\Delta t}{2C_{\text{stub}}} \tag{23}$$

Burfoot and James[5] used a finite difference technique to model heat flow in a sphere of lean meat. Figure 15 shows a comparison between results obtained by TLM modelling and those of Burfoot and James. The continuous line represents the results obtained by the finite difference approach, while the TLM results are represented by solid circles. The TLM model used exactly the same size of sphere and values of specific heat, density and thermal conductivity as the finite difference model. It is also limited by the same assumptions, namely:

1 That the atmosphere is unaffected by the presence of the meat sphere, and that exposure of the meat to the heating atmosphere is instantaneous.

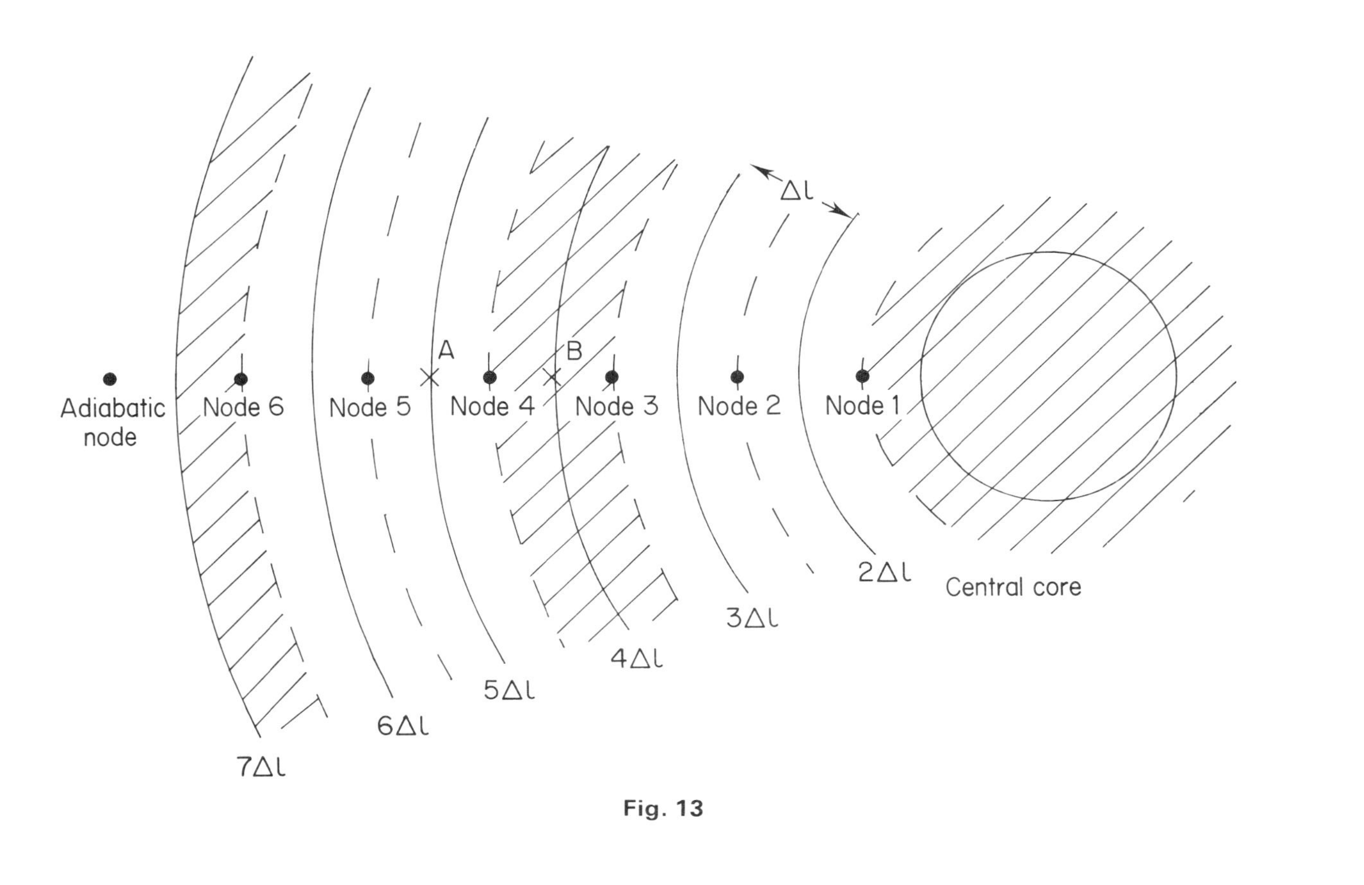

Δl
Adiabatic node
Node 6
Node 5
A
Node 4
B
Node 3
Node 2
Node 1
Central core
2Δl
3Δl
4Δl
5Δl
6Δl
7Δl

Fig. 13

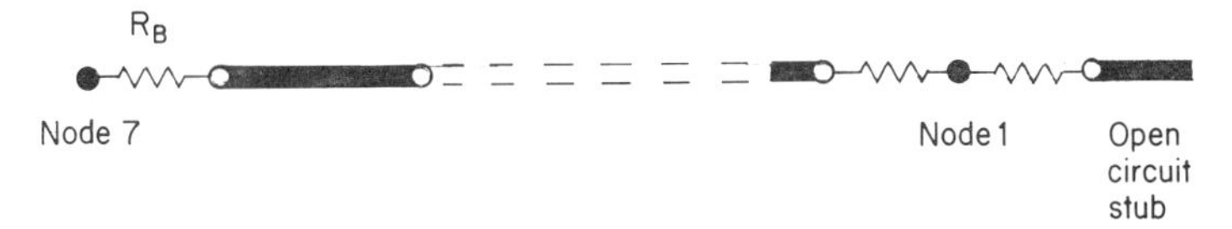

Fig. 14

2 That the sphere is homogeneous and its thermal properties constant with temperature.
3 That the dimensions of the sphere are constant throughout.

As $R \sim 1/\Delta l$ and $C \sim \Delta l^3$, the speed at which the problem moves varies as the square of the mesh size. Even so, using a mesh size of 0.01 m to model a sphere of radius 0.07 m, a timestep of 2 s gave no obvious oscillations and produced results comparable with those obtained by a finite difference routine, as illustrated in Fig. 15.

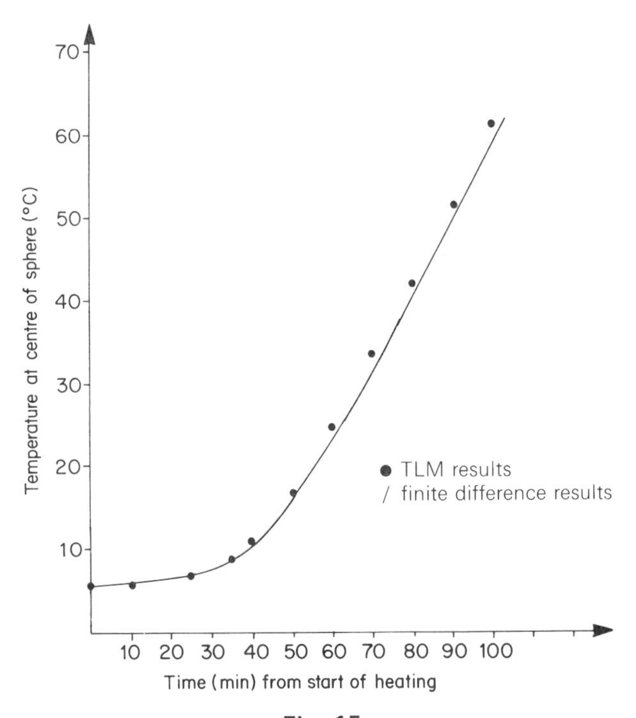

Fig. 15

B Modelling in two and three dimensions

The majority of modelling situations demand consideration in two or three dimensions. This section describes modelling, in two dimensions, of a non-homogeneous, irregularly shaped, meat joint.

The geometry of an irregular body must be discretized before modelling can be attempted. Figure 16 illustrates the discretization process for the joint in question. The cross-section is first overlaid by a square Cartesian mesh and its geometry is approximated to the mesh. Each mesh square is considered to contain only one material or to fulfil only one role—in this example the regions marked ● were designated as boundary regions and were characterized by a boundary resistance, R_B, which was calculated from equation 22. Clearly the mesh size governs the accuracy of this discretization process.

In this example, nodes in the TLM network were taken to be at the centres of mesh squares, and each node was considered to be surrounded by four identical resistors whose values were derived from equation 5. From the point of view of constructing a scattering matrix in two-dimensions and operation of the TLM algorithm it is also desirable to have not only identical resistors clustered around any node but also link transmission lines of identical impedance connecting nodes to each other. In order to facilitate this the capacitance values modelled in the link lines throughout the transmission line network were the minimum appropriate to any part of the network. Additional heat capacity in any region was then modelled by a stub transmission line connected directly to the node. Figure 17 illustrates the scheme for a single node, the stub resistance R_S being included only for completeness.

Applying the circuit theorems of Norton and Thevenin to a single node gives the following relationships for reflected pulses and nodal temperatures

$$T = \left[\frac{2}{(R + Z)} \sum_{l=1}^{l=4} V_l^i + \frac{2}{(R_s + Z_s)} V_s^i \right] \tag{24}$$

$$V_l^r = \frac{Z}{(R + Z)} T + \frac{(R - Z)}{(R + Z)} V_l^i \tag{25}$$

$$V_s^r = \frac{Z_s T}{(R_s + Z_s)} + \frac{(R_s - Z_s)}{(R_s + Z_s)} V_s^i \tag{26}$$

where:

T ≡ nodal temperature
V_l^i, V_l^r ≡ incident and reflected pulses on the link lines
V_s^i, V_s^r ≡ incident and reflected pulses on the stub line
R_s, Z_s ≡ resistance and impedance of the stub

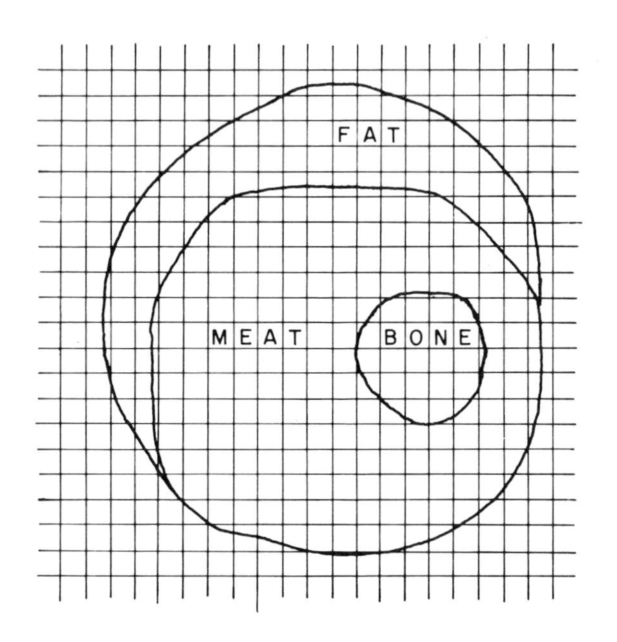
FAT
MEAT
BONE

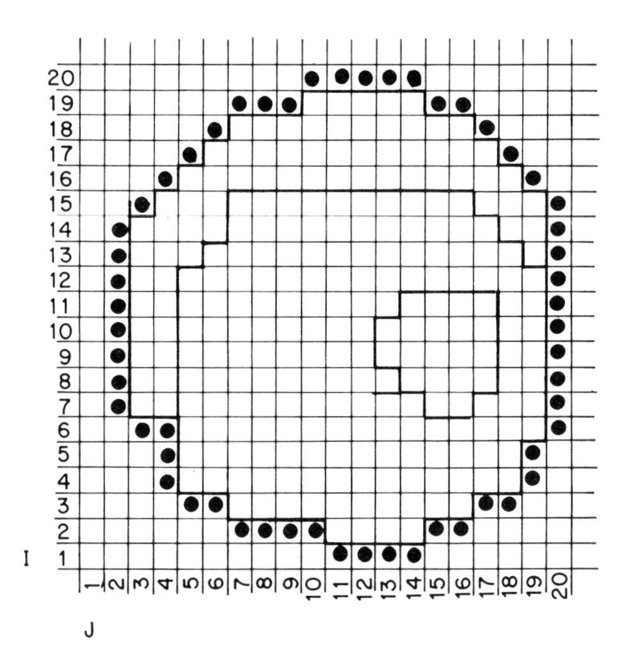
20
19
18
17
16
15
14
13
12
11
10
9
8
7
6
5
4
3
2
1
I
1 2 3 4 5 6 7 8 9 10 11 12 13 14 15 16 17 18 19 20
J

Fig. 16

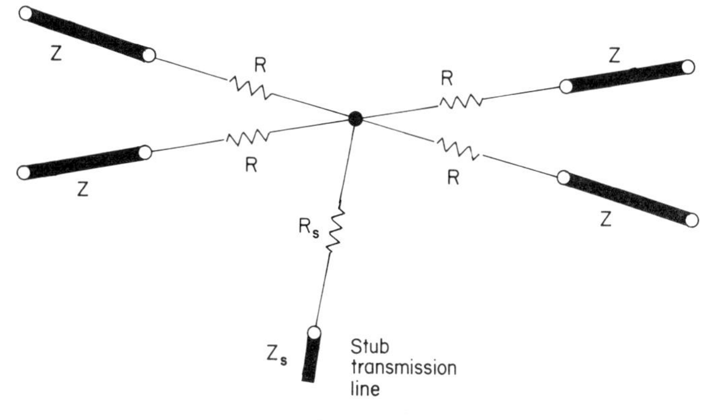

Fig. 17

Although these relationships were derived using circuit theorems, considerations similar to those outlined in section 2 can be used to show that they are reasonable in terms of heat flow.

Figure 18 shows the temperature distribution through the row of nodes at $I = 10$ in Fig. 16. The values of the thermal parameters used in the computation were:

$$K_{\text{meat}} = 0.08 + 0.52\,X_w \text{ W K}^{-1}\text{ m}^{-1}$$
$$S_{\text{meat}} = 1672 + 2508\,X_w \text{ J kg}^{-1}\text{ K}^{-1}$$
$$\rho_{\text{meat}} = 890 + 100\,X_w \text{ kg m}^{-3}$$
$$K_{\text{fat}} = (0.08 + 0.52\,X_w)/2 \text{ W K}^{-1}\text{ m}^{-1}$$
$$S_{\text{fat}} = 1672 + 2508\,X_w \text{ J kg}^{-1}\text{ K}^{-1}$$
$$\rho_{\text{fat}} = (890 + 110\,X_w)/2 \text{ kg m}^{-3}$$
$$K_{\text{bone}} = 2\,(0.08 + 0.52\,X_w) \text{ W K}^{-1}\text{ m}^{-1}$$
$$S_{\text{bone}} = 2\,(1672 + 2508\,X_w)/2 \text{ J kg}^{-1}\text{ K}^{-1}$$
$$\rho_{\text{bone}} = (890 + 110\,X_w)/2 \text{ kg m}^{-3}$$

$h = 18.5 \text{ W m}^{-2}\text{ K}^{-1}$, X_w being the % water content normalized to one.

Calculations were based on an air temperature of 200 °C. The mesh size, Δl, was 0.01 m throughout and a timestep of 4 s did not result in obvious oscillations. The meat, initially at 0 °C but unfrozen, was subjected instantaneously to an atmosphere at 200 °C.

The model described in this section could only realistically be applied to a joint of constant cross-section whose length was sufficient to render the effect of heat input through the ends negligible. Other, more complex, situations demand modelling in three-dimensions but this poses no great problems.

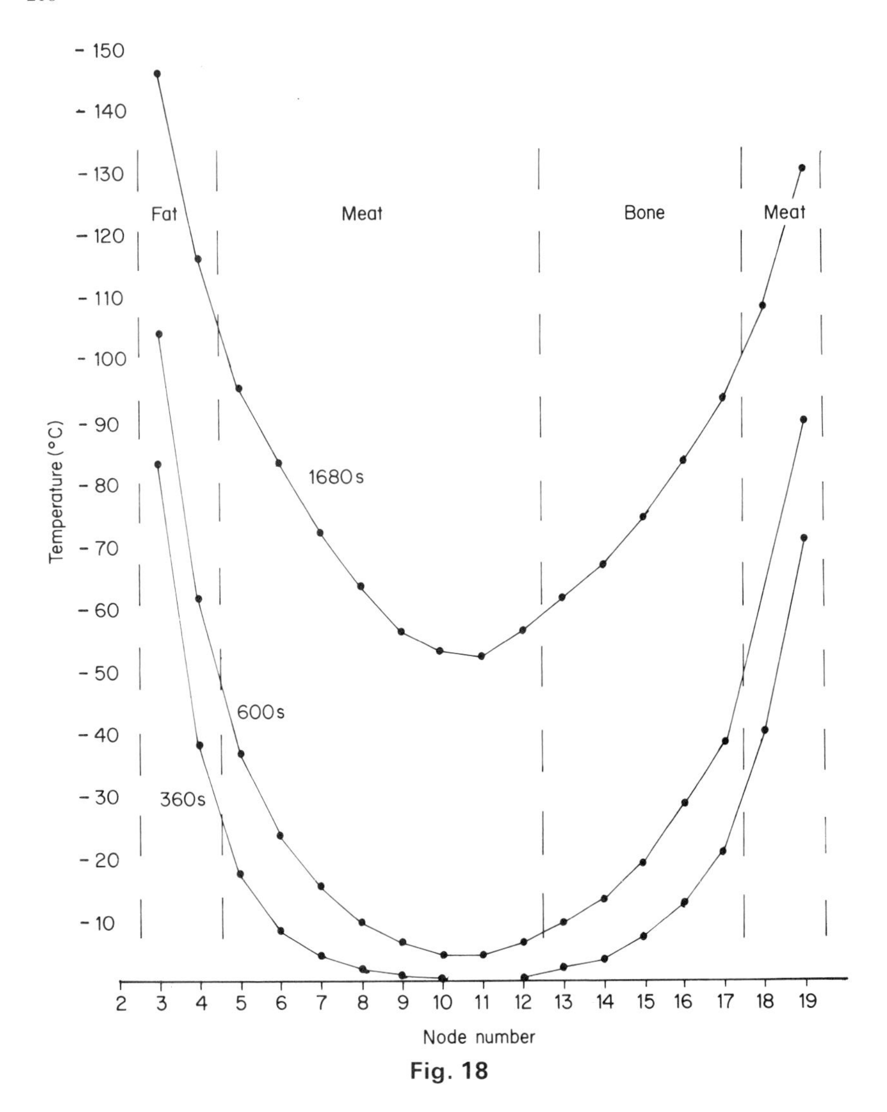

Fig. 18

References

1. Johns, P. B. (1977). *Int. J. Num. Meth. in Eng.* **11**, 1307–1328.
2. Johns, P. B. and Butler, G. (1983). *Int. J. Num. Meth. in Eng.* **19**, 1549–1554.
3. Johns, P. B. and Butler, G. (1979). Proc. of the First International Conf. held at University College, Swansea, July 1979.
4. Lohse, A., Johns, P. B. and Wexler, A. (1977). *IEEE Transactions on Education* **20**, 64–68.
5. Burfoot, D. and James, S. J. (1983). COST 91 Proceedings, Athens, November 1983.

11 The Nature and Significance of Heat and Mass Transfer Processes in Foodstuffs

D. BURFOOT

Meat Research Institute, Bristol

1 INTRODUCTION

The transfer of heat and mass both within, and to and from foodstuffs, forms the basis of the majority of food processing operations. Most foods pass through a series, or chain, of processing operations prior to consumption including, for example, slaughter or harvesting, chilling, chilled storage and cooking. The range of processes varies between foodstuffs and the properties required of the final product. If, for example, longer storage lives are required, foods are frozen, sterilized, pasteurized, canned, dried or cured, all of which involve heat and/or mass transfer.

In all these processes and with many others the aim is to increase or retain the nutritional, sensory or economic value of the food. Often the requirements conflict and the choice of operating conditions is not simple but is

FOOD STRUCTURE AND BEHAVIOUR
ISBN 0-12-104230-8

usually based on a combination of empirical and fundamental design equations.

A simple example of chilling meat carcasses illustrates the role of heat and mass transfer and some of the conflicting requirements. Immediately after slaughter and dressing a beef carcass is hot (approximately 40 °C) and wet, and left in this state the meat surface provides ideal conditions for microbial growth and a high rate of evaporative weight loss. It is also impossible to conventionally cut and joint meat at this temperature, hence the carcasses must be cooled and in the UK this is normally carried out in chill rooms using moving, refrigerated air. Problems arise in specifying the optimum environmental conditions for the cooling process, e.g. air temperature, velocity and humidity. There are conflicting requirements to satisfy microbiological aims, which include producing a dry surface, and those of the meat trade which tries to minimize weight loss.

If instead of a carcass we were cooling a steel bar the thermal properties would relatively be more certain and solving the heat conduction equation would indicate the cooling time between two temperatures that could be achieved with various air conditions. If we return to the problems with beef carcasses, an important feature to note is that they are of irregular shapes (no two are identical) and are inhomogeneous, comprising fat, lean and bone in varying proportions. A cursory study of the data on their thermophysical properties clearly shows that not only are there considerable differences between the major components but that the properties of fat and lean vary substantially with temperature. A further difference between cooling foods and steel is that while both will lose heat by conduction, convection and radiation, foods also lose heat by evaporation of water from their surfaces. To add further complications, using low air temperatures (below −1 °C) for extended periods can produce undesirable surface freezing and, with meat, very-high cooling rates may lead to irreversible textural changes.

With such conflicting considerations no single set of conditions will provide an ideal solution. However, an understanding of the concepts and equations that underlie the transport phenomena can help to optimize the conditions in each situation. In this paper these concepts are introduced, discussed and illustrated with typical applications.

2 HEAT TRANSFER

Heat transfer without a phase change is the flow of energy down a temperature gradient and occurs by three methods: conduction, convection and radiation. Examples given in this section show that in many practical situations more than one transfer mode is important. However, for clarity, the principal equations and mechanisms describing each of these modes are considered separately.

A Conduction

The conduction of heat occurs by virtue of the transfer of kinetic energy between the vibrating molecules of a substance. The rate of one dimensional heat transfer by this mechanism is given by:

$$q_{\text{cond}} = kA \frac{\partial T}{\partial x} \tag{1}$$

where q_{cond} = rate of heat transfer by conduction, k = thermal conductivity of the body through which heat is transferring, A = cross-sectional area of the body, T = temperature and x = distance.

Using this equation and applying a heat balance over a small time increment within a small segment of the body, it can be shown that:

$$\frac{\partial T}{\partial t} = \frac{k}{c\rho}\left(\frac{\partial^2 T}{\partial x^2} + \frac{a}{x}\frac{\partial T}{\partial x}\right) \tag{2}$$

where t = time, c = specific heat capacity, ρ = density and a = geometric factor; for a slab, $a = 0$; for a cylinder, $a = 1$; for a sphere, $a = 2$.

Equation 2 describes the time dependent variation of temperature within a sphere, or an infinite cylinder or slab. Before the temperature-time relationship can be calculated, it is necessary to impose certain restrictions on the equation. These boundary conditions usually consist of a heat balance at the surface of the body and descriptions of the initial conditions and any geometrical symmetry: a typical description is:

$$t > 0: \quad x = X; \quad k\frac{\partial T}{\partial x} = \text{heat flux to or from the surface} \tag{3}$$

$$t = 0; \quad x < X; \quad T = T_i \tag{4}$$

$$t \geqslant 0; \quad x = 0; \quad \frac{\partial T}{\partial x} = 0 \tag{5}$$

where X = radius or half thickness of the body, T_i = initial uniform temperature and the mathematical descriptions of the heat flow to or from the surface will be derived later, see for example equation 12.

Graphical representations of the solutions to equations 2 to 5 can be found in many textbooks of heat transfer and food engineering;[19,28,35] the solutions, and those for bricks and finite cylinders, are presented in a convenient form by Dalgleish and Ede[16] and approximate solutions by

Ramaswamy *et al.*[44] All of these solutions were derived using analytical techniques to solve the conduction equation. Unfortunately, with these methods it is usual to consider the particular case of constant material properties, dimensions and heat transfer coefficient, which rarely applies to foodstuffs. These restrictions can be overcome by the use of numerical finite difference and finite element techniques and the increasing availability of computers has led to a large number of publications describing these techniques. A description of these methods is beyond the scope of this discussion but typical applications include freezing,[9-11] freeze drying,[34] thawing,[11,21] cooling[1] and cooking[8,14,15,20,52] Krishna Murthy and Badari Narayana[33] consider many of these processes and drying. A further analysis technique, the transmission line matrix method, is considered by Johns.[24]

Data on the properties of foodstuffs which are required for the numerical analyses may be obtained from the references cited by Adam[2] and Krishna Murthy and Badari Narayana.[33] Further data for particular foods include potatoes,[57] meat[3,38] and fish.[23,45] The compilation of Miles *et al.*[37] is useful over the temperature range of $-40\,°C$–$+40\,°C$ but, as emphasized by the authors, the data apply to atmospheric conditions with foods that have not been stored for long periods and have not undergone any significant heat treatment. As illustrated by EEC collaborative work,[25] these limitations and complications such as the variability within foodstuffs and the difficulties of measuring their properties, lead to uncertainties in design calculations. Nonetheless, it can be concluded that the thermal conductivities and thermal diffusivities of most foodstuffs are sufficiently low for the rate of temperature

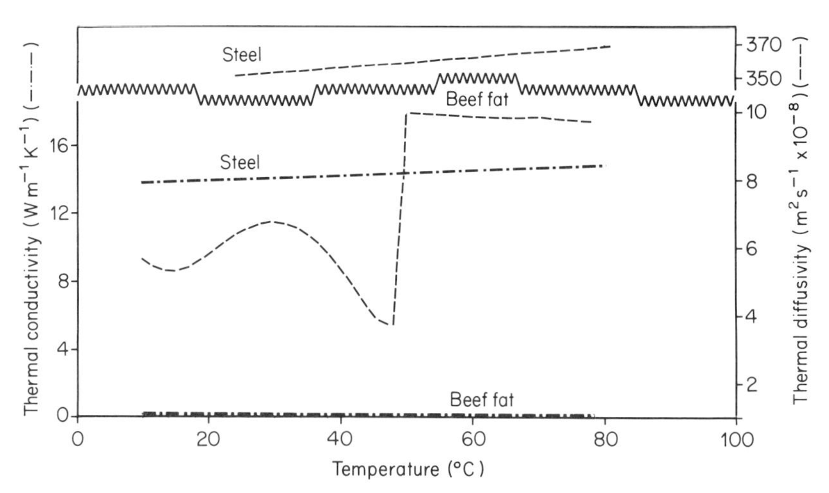

Fig. 1 Comparison of the thermal properties of beef fat[3,37] and steel

rise within the material to be limited by slow internal diffusion. An example of a diffusion controlled series resistance process occurs during the heat sterilization of canned products in which it is relatively easy to transfer heat to the surface of the food, but considerably more difficult to conduct it through the material to the centre. Similar difficulties arise during the roasting of potatoes in a domestic oven, but in this process the use of metal skewers, with a high thermal conductivity, can assist the conduction of heat. Figure 1 illustrates a particularly marked contrast between the properties of a foodstuff, beef fat, and a common engineering material, steel. Also shown is the significant temperature dependency of the food properties which invalidates the constant property assumption of many analytical models.

B Convection

The exchange of heat between a body and any surrounding fluid may be aided by the motion of that fluid. This transfer of heat is called "free or natural convection" or "forced convection". In the former case the fluid motion is due to the density variations caused by the heat transfer and in the "forced" example the motion is caused by an external mover such as a fan. The rate of heat transfer by either of the convection mechanisms is described by:

$$q_c = h_c A \, \Delta T \tag{6}$$

where q_c = rate of heat transfer by convection, h_c = convection, or film, heat transfer coefficient and ΔT = difference between the temperatures of the bulk of the fluid and the surface of the body.

It is useful to consider that all of the temperature difference, $\Delta T = (T_a - T_s)$, exists over a layer of fluid near to the surface of the body and there is no temperature gradient within the well mixed bulk of the fluid. Hence all of the resistance to heat transfer lies in the hypothetical thermal boundary layer of thickness, δ (Fig. 2). Assuming that heat is transferred purely by

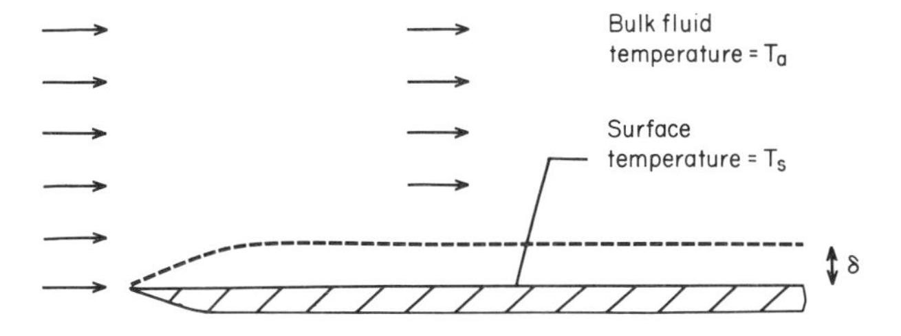

Fig. 2 The concept of a thermal boundary layer

conduction through this layer, it follows from equations 1 and 6 that:

$$h_c = \frac{k}{\delta} \tag{7}$$

where k = thermal conductivity of the fluid.

It is clear from equation 7 that reducing the thickness of the thermal boundary layer produces larger heat transfer coefficients and this can be achieved by, for example, increasing the fluid velocity. The higher fluid velocities generally associated with forced, as opposed to free, convection result in significantly larger transfer coefficients. Consequently, the rates of heat transfer, and hence product throughput, may be increased by inducing circulation around a foodstuff. This usually increases the complexity and capital cost of the equipment which is probably why domestic appliances, e.g. home freezers and many ovens, unlike their industrial counterparts, do not incorporate fans.

Equation 7 provides a useful representation of the mechanism of convective heat transfer and the use of the boundary layer concept can lead to equations for predicting heat transfer coefficients.[12,47] Unfortunately, the effective thickness, δ, is unknown in most practical situations and the techniques of dimensional analysis and experimentation are used. These indicate that for convective heat transfer:[13]

$$Nu = f(Re, Pr, Gr) \tag{8}$$

where

$$Nu = \text{Nusselt number} = \frac{\text{heat transfer rate}}{\text{rate of conduction through the fluid}}$$

$$= \frac{h_c A\,\Delta T}{kA\,\Delta T/L} = \frac{h_c L}{k}$$

$$Re = \text{Reynolds number} = \frac{\rho u^2}{\mu u/L} = \frac{Lu\rho}{\mu}$$

$$Pr = \text{Prandtl number} = \frac{\text{kinematic viscosity}}{\text{thermal diffusivity}} = \frac{\mu c \rho}{\rho k} = \frac{c\mu}{k}$$

$$Gr = \text{Grashof number} = \frac{\beta g\,\Delta T L^3 \rho^2}{\mu^2}$$

L = characteristic dimension, for example, the inside diameter of a tube, u, ρ, μ, β = respectively velocity, density, viscosity and coefficient of thermal expansion of the fluid and g = gravitational constant = 9.81 m s^{-2}.

Data obtained from experiments designed to measure heat transfer coefficients are correlated using various forms of equation 8. (For each correlation it is important to note the definition of the heat transfer coefficient, h_c, in many cases, it is based on a temperature difference between the bulk mixed mean and surface temperatures). The most common form of equation 8 is:

$$Nu = aRe^b Pr^c Gr^e \tag{9}$$

Values of the parameters, a, b, c and e depend on the significance and magnitude of the dimensionless numbers. The Reynolds and Grashof numbers are respectively measures of the ratio of the inertial to viscous forces and the ratio of buoyancy to viscous forces. During natural convection when inertial forces are small and buoyancy and viscous forces are significant ($c, e > 0$), the effect of Reynolds number is negligible ($b = 0$), while during forced convection the converse applies ($b, c > 0, e = 0$). Values of a, b, c and e are presented by Perry and Chilton[39] for various ranges of the dimensionless groups. A particularly notable and practically important change of the parameters occurs at Reynolds numbers of approximately 2100 for pipe flow and 10^5 for external flows. Below these values an element of fluid, or a small particle in the fluid, would follow a straight path parallel to the surface and the flow is termed "laminar". At higher fluid velocities or lower viscosities the path of the element would be less well defined. The random motion and mixing which then occurs within this "turbulent" fluid produces comparatively higher heat transfer coefficients.

Further complications in estimating heat transfer coefficients are caused by the dimensions of the heat transfer surface and the direction of heat flow. For some distance along the surface, known as the calming length, the fluid undergoes a transformation as the flow pattern is stabilized. Along this region the boundary layer is developing and consequently the heat transfer coefficients are larger than elsewhere. During heating the lower viscosity of the fluid near the heating surface results in a thinner boundary layer and larger heat transfer coefficients than during cooling. This phenomenon is particularly important when fluid properties are highly-temperature dependent, as with many foods. For pipe flow these complications are considered by including in the correlating equation the ratio (d/l) and the ratio of the viscosities of the bulk fluid and at the tube wall (μ/μ_w), e.g.

$$Nu = 1.86\left(RePr\frac{d}{l}\right)^{0.33}\left(\frac{\mu}{\mu_w}\right)^{0.14} \quad \text{for} \quad \frac{du\rho}{\mu} < 2100 \tag{10}$$

$$Nu = 0.023Re^{0.8}Pr^{0.33}\left(\frac{\mu}{\mu_w}\right)^{0.14}\left[1 + \gamma\left(\frac{d}{l}\right)\right] \quad \text{for} \quad \frac{du\rho}{\mu} > 2100 \tag{11}$$

where the parameter γ depends on the geometry of the pipe inlet, the length of pipe and the Reynolds number.[27,32,40]

These correlating equations containing heat transfer coefficients are simple forms used to represent the effects of phenomena about which little is known. Many studies have been, and still are, devoted to the measurement and understanding of fluid flow and heat transfer. Most of this work applies to simple geometries and Newtonian fluids and there is a particular dearth of understanding of external flows around irregular and variable objects such as animal carcasses and vegetables. Studies have been undertaken to produce correlations applicable to flows in jacketed vessels with or without stirrers and coils.[55] Skelland[49] presented many of the available correlations for designs with non-Newtonian fluids and showed that most of the correlations considered above may be applied provided that suitable fluid properties are chosen. The measurement of those properties is inherently difficult[43] and there is a serious lack of heat transfer correlations and thermophysical data for food slurries.

Before considering other mechanisms of heat transfer, two examples of the significance of heat transfer coefficients will be investigated.

Equation 3 may be rewritten to include the effect of purely convective heat transfer to a surface:

$$k \frac{\partial T}{\partial x} = h_c \, \Delta T \tag{12}$$

Solving equations 2, 4, 5 and 12 shows that for Biot numbers ($Bi = hX/k$) less than 0.1, the transfer of heat is controlled by the external resistance. Under these conditions, increasing the heat transfer coefficient will reduce the processing times. With meat chilling this may cause "cold shortening", a condition manifest as toughness. In other processes there might be no such problem. For Biot numbers greater than 40, the heat transfer rate is controlled by the "internal resistance" of the food. Operating in this range may lead to surface degradation, such as burning during cooking, and significant losses of moisture. However, these conditions may be acceptable, for example during grilling. As a general guide, although there are a lot of exceptions, many processes are operated in the intermediate Biot number range i.e. $0.1 < Bi < 40$.

Another example of series resistances, where considerable advantages may result from the correct choice of heat transfer coefficients, occurs in the design of a heat exchanger. Under steady state conditions, the rate of heat transfer between two fluids in a double pipe exchanger may be calculated using:

$$q = UA \, \Delta T_{LM} \tag{13}$$

where

$$\frac{1}{U} = \frac{1}{h_{co}} + \frac{x_{Ro}}{k_{Ro}} + \frac{x_w d_O}{k_w d_{LM}} + \frac{x_{Ri}}{k_{Ri}} + \frac{d_O}{dh_c} \tag{14}$$

i.e. Overall resistance to heat transfer = sum of the resistances in the outside fluid film, outside wall scale, tube wall, inside wall scale and inside fluid film

where h_c, h_{co} = heat transfer coefficients acting inside and outside of the tube, d, d_o = inside and outside diameters of the tube,

d_{LM} = logarithmic mean diameter of the tube,

$= (d_0 - d)/\ln(d_0/d)$

k_{Ri}, k_{Ro}, k_w = thermal conductivities of the scale on the inside and outside surfaces of the tube, and of the tube wall, x_{Ri}, x_{Ro}, x_w = thickness of the layer of scale inside and outside the tube, and of the tube wall.

These equations are derived by applying equation 1 to the tube wall and the scale present on both sides of the wall, and equation 6 to the fluid layers adjacent to the pipe.[29] The scale resistances are usually determined from previous experience[30] and denoted by the symbols Ri and Ro. The logarithmic mean temperature difference, ΔT_{LM}, is a measure of the effective temperature difference between the fluids.

Equation 14 shows that the overall resistance to heat transfer, $1/U$, is equal to the sum of the individual resistances acting in series. This important concept, which appears again in the discussion of mass transfer, shows that a reduction of the resistance to heat transfer is best achieved by reducing the dominant factor in equation 14. The largest resistance lies in the fluid food and although it can be reduced by increasing the fluid velocity, such an increase is limited by the resulting degradation of the food. Novel devices that agitate the fluid, and thereby disrupt the temperature profile and the boundary layer, have been developed to alleviate this problem which is prevalent in viscous and non-Newtonian fluids. Typical examples of such equipment include jacketed inline mixers,[7] scraped surface heat exchangers and hollow cut flight mixers.[55] Overall heat transfer coefficients for these devices operating under typical conditions can be found in the manufacturer's literature. These values, when substituted into equation 13, are useful during the early stages of plant design but recourse to equation 14 shows that individual U-values are applicable to limited conditions and the manufacturer's suggestions should be checked at a later design stage.

C Radiation

Thermal electromagnetic radiation is believed to result from the motion of the atoms and molecules within an object. These motions increase with temperature and the associated emitted radiation may be partially absorbed by any object upon which the radiation is incident. The resultant temperature rise of the second object demonstrates that heat is transferred. The rate of transfer may be estimated using:

$$q_r = F_A F_\varepsilon A_2 \sigma (T_1^4 - T_2^4) \tag{15}$$

where q_r = rate of heat transfer by radiation, A_2 = surface area of body 2, T_1, T_2 = absolute temperature of bodies 1 and 2, and σ = Stefan-Boltzmann constant. The geometry factor, F_A, and the emissivity factor, F_ε, have been calculated for some common situations.[31] However, in many situations the emissivity of the walls of the processing vessel are unknown, but with a small food sample in a large enclosure the following approximation may be applied:

$$q_r = \varepsilon A_2 \sigma (T_1^4 - T_2^4) \tag{16}$$

Values of the emissivities, ε, of some foodstuffs are given by Miles.[36]

The dependence of the transfer rate on the fourth power of the temperatures indicates that radiation phenomena are important when large temperature differences are used. In particular, this is true of heating processes such as cooking. In the initial stages of cooking a meat joint in a natural convection oven at 175 °C the rate of heat input by radiation is approximately 1.25 times that by convection. Although the charts considered earlier for the prediction of heating times do not immediately appear to be suitable for use with equation 16 this problem can be overcome by defining a radiation transfer coefficient, h_r, such that:

$$q_r = h_r A (T_1 - T_2) \tag{17}$$

The application of this equation is restricted to systems where no radiation returns to the body from which it was emitted. In this case, equation 16 can be combined with equation 17 to estimate the radiation coefficients which can be used for the prediction of heating times in simple situations.

D Other mechanisms of heating

Heating by the mechanisms discussed already relies on the existence of a temperature gradient; the direction of which determines whether the heat flows into or out of the food. The use of large temperature differences results in short processing times but the foodstuff may be degraded. This problem has stimulated a substantial research effort investigating techniques that do not rely on temperature differences, but apply instead electromagnetic waves which rotate the dipoles of the molecules within the food, the resultant friction being converted into heat. Microwave and capacitive heating are two techniques that rely on this phenomenon, although there does not appear to be an analogue of this mechanism that would facilitate the extraction of heat. In microwave heating, electromagnetic waves with a frequency of 896, 915 or

2450 MHz are directed towards the food by waveguides, whereas for capacitive heating, electrical energy, usually at frequencies of 13.6, 27.1 or 40.7 MHz, is applied to the food via two electrodes. The heating capabilities of both techniques are described by:

$$P/V = 15.6 \times 10^{-14} E^2 \omega \varepsilon_r \quad (18)$$

where P/V = power per unit volume of food, W cm^{-3}, E = field strength, V cm^{-1}, ω = frequency, Hz, and ε_r = effective loss factor which depends on the dielectric properties of the material.

The applications and difficulties of using electromagnetic heating are extensively considered by Smith.[50,51] Successful applications include tempering of frozen meat, reheating of many foods, precooking of bacon slices and proofing of doughnuts. Although some sales literature implies that electromagnetic heating produces a uniform temperature distribution within foods, this is not achieved in practice. However, the maldistribution of the high energy density is beneficial for rapid drying, for example of pasta.

There are several problems associated with electromagnetic heating, they include:

i Heating rates are not uniform because of the irregular shape of many foods, the difficulty of producing a uniform energy field and the dependency of the dielectric properties on the temperature and composition of the food and the frequency of the radiation. A good example of the difficulties is the microwave thawing of meat. The loss factor for water is greater than for ice so that, after the initial thawing at the surface, energy is preferentially dissipated at the surface of the meat causing drying and sometimes partial cooking which reduce the value of the meat.

ii The capital cost of equipment is very large compared to most conventional heating processes and this can limit electromagnetic heating applications to high value products and continuous plant operation. For example, the blanching of vegetables may be achieved using microwaves but the relatively short processing periods during a year hardly justify the capital expenditure.

iii The cooking treatments required by many foods depend upon an integration of processing time and temperature and although electromagnetic heating can rapidly increase the temperature, it does not have the tenderizing properties of prolonged heating.

iv The frequencies assigned to electromagnetic radiation for heating applications will not permit penetration through metals, hence canned foods cannot be heated.

Some of these problems can be alleviated using combinations of electromagnetic and conventional heating which also reduces the amount of expensive electromagnetic heating that is required. Good examples of such applications include baking bread followed by surface browning by hot air and drying operations with hot air used during the initial stages.

Applications of electromagnetic heating have been considered in detail. The nature of heat flow by other mechanisms has been shown with reference to the phenomena, governing equations and practical examples. The significance of these and other modes of heat flow is also related to the mass transfer that occurs simultaneously and is affected by similar parameters such as the air velocity. Discussion is now directed towards these modes of mass transfer.

3 MASS TRANSFER

The term "mass transfer" refers to the movement of a material down a concentration gradient or between phases. The following sections illustrate that many of these transfers are analogous to some of the modes of heat transfer and similar forms of correlation and equations are used in both cases.

A Diffusion

The rate of transfer of a substance, B, by diffusion through a solid, C, is governed by:

$$N_B = D_{BC} A \frac{\partial c_B}{\partial x} \tag{19}$$

where N_B = rate of mass transfer, D = diffusion coefficient, or mass diffusivity, of B through C and c_B = concentration of B. A comparison of equations 1 and 19 reveals that the transfer rates of heat and mass within a body are proportional to the cross-sectional area of the body and a driving force which, in the respective examples, is the temperature gradient ($\partial T/\partial x$) or the concentration gradient ($\partial c/\partial x$). The two factors of proportionality are the thermal conductivity and the mass diffusivity. The assumption in both equations is that the transfers occur by only one mechanism, heat conduction or mass diffusion. Hence, without modification, no allowance is made for other transfer mechanisms such as bulk flow of B through C as a result of a total pressure difference.

Using the same method of derivation as for equation 2 it is found that the variation of concentration with respect to time is governed by:

$$\frac{\partial c_B}{\partial t} = D_{BC}\left(\frac{\partial^2 c_B}{\partial x^2} + \frac{a}{x}\frac{\partial c_B}{\partial x}\right) \tag{20}$$

where a = geometric factor (as for equation 2).

A typical set of boundary conditions is derived by replacing the temperature and conductivity in equations 3 to 5 with concentration and mass diffusivity. Equation 3 would then be replaced by:

$$D_{BC}\frac{\partial c_B}{\partial x} = \text{mass flux of substance } B \text{ to or from the surface} \qquad (21)$$

Since equivalent equation structures govern the internal transfers of heat and mass, identical methods are used to solve both sets of equations. It must be stressed again that in using equation 19 it is inherently assumed that the transfer of mass occurs purely by diffusion. Consider, for example, two cases for which this assumption is invalid. During the cooking of meat, or indeed many foods, water diffuses outwards to replace the loss of water at the surface. As the meat temperature increases, proteins denature causing shrinkage of the solids network and consequently water is forced out through the meat. Most shrinkage occurs at temperatures exceeding 50 °C so that although the diffusion equation may apply up to that temperature, thereafter allowance must be made for the simultaneous movements of "cooking drip" which are not the result of a concentration gradient (diffusion). The leaching of sugar beet cossettes provides a further example of the difficulties of simulating mass transfer within foodstuffs. In this process a liquid solvent, water at approximately 75 °C, is passed through a bed of the sugar beet slices with the aim of selectively dissolving the sugar. This appears to be a diffusion process and certainly is within the slices, however, the slicing operation damages many of the sugar beet cells such that a significant proportion of the sugar is recovered by washing rather than leaching.

During the later stages of some food drying operations it is comparatively easy, by using high air velocities, to transfer moisture away from the surface of the food into the air. The rate of moisture removal is then controlled by internal diffusion and this is clearly similar to the thermal diffusion control during sterilization. In drying terminology, the period of diffusion control is often termed the "falling rate period" because the rate of moisture removal decreases as a result of the reduced concentration gradients within the food. Jason[22] found that two falling rate periods occur during the drying of fish muscle. The first period is characterized by a higher diffusion coefficient and it is proposed by Jason that the transfer to the second period is associated with the removal of a unimolecular layer of water attached to the solids. The difficulty of detaching this "strongly bound" water is the cause of the lower diffusion coefficient in the second falling rate period. Although few complete studies of animal tissue drying have been published, it is unlikely that the existence of more than one falling rate period is peculiar to fish.

Mechanical methods are often used to reduce the size of foodstuffs which eases the mixing processes and decreases the diffusion path of additives through the food. For example, sodium chloride is often added to improve taste, increase storage life and disrupt the cells of the food: citric and tartaric acids are added to some processed meats to improve the water holding capacity: monosodium glutamate is used to enhance flavour, and nitrites, nitrates and ascorbic acid may be added to produce and retain a desired colour.[48] In each of these examples the additives will diffuse into the food particles and increase their functional ability. Reducing the food particle size is not always an economical method of decreasing the diffusion path, particularly with high value foods such as prime meat. Multi-needle injection of brine is often used to cure meat and this reduces the diffusion path within the meat but subsequent storage in a brine tank for 5 days followed by 9 days in air is required to allow sufficient time for the diffusion process. This processing time is long compared to most other operations and requires significant floor space. To enhance the diffusion rate, one process (vacuum curing) involves multi-needle injection followed by continuous cycling of the meat under vacuum and atmospheric pressures. This causes the meat fibres to repeatedly separate and move together aiding the penetration of brine.

In most of the above examples, the diffusing substance moves along the pores of the solid matrix and not along the normal diffusion path; it is usual therefore to refer to D_{BC} as the "effective" or "apparent" diffusivity. This term is also used to define the diffusivity of a fluid substance in a fluid mixture.[6] In cases of diffusing fluids, equations 19 and 20 are applicable. For gases, it is customary to consider partial pressures, rather than concentrations, because these can be measured more easily. Experimental diffusivity values should be used when available. In other instances, often in the early stages of plant design, recourse to the kinetic theory of gases may provide adequate estimates of the diffusivities.[41] This approach reveals that gas diffusivities are dependent on the substances involved and the temperature and pressure of the system. Typical accuracies of the estimated D values are 8% but errors up to 25% may occur for polar molecules. The lack of a good kinetic theory for liquids indicates that this approach cannot be adopted for predicting liquid diffusivities and the available correlations[41] are less reliable than for gases, errors being up to 50% in the worst cases. Unfortunately, experimental values of liquid diffusivities are limited to dilute solutions, atmospheric pressure and the temperature range of 0–40 °C. Values for the concentrated liquids found in many food operations are difficult to locate.

B Convection

The diffusion equation can be used to estimate the rate of transfer of a substance through any phase. However, such estimates are only valid if the phases are stagnant or moving slowly (laminar flow), otherwise the substance is transferred within each phase by a mixing phenomenon. Diffusion is often a slow process and because commercial operations require high throughputs, it is usual to induce mixing. At the molecular level, mixing is a diffusion process, but it is difficult to consider it mathematically in this way. For this reason, the transfer rate is estimated using a different approach. For a liquid/gas system we may use:

$$N_B = h_{mL} A \, \Delta c_B \tag{22}$$

or

$$N_B = h_{mG} A \, \Delta p_B \tag{23}$$

where h_{mL} = liquid side mass transfer coefficient and Δc_B = difference of concentrations of B within the bulk of the liquid and at the liquid/gas interphase, h_{mG} = gas side mass transfer coefficient and Δp_B = partial pressure difference of B within the bulk of the gas and at the interphase.

The mixing process within each phase may be enhanced by using high fluid velocities but even with good mixing, relatively stagnant layers of fluid will exist adjacent to both sides of the liquid/gas interface. If it is assumed that all of the mass transfer driving force (concentration or partial pressure difference) exists across these two layers then:

$$h_{mL} = \frac{D_{BL}}{\delta_{mL}} \tag{24}$$

and

$$h_{mG} = \frac{D_{BG}}{\delta_{mG}} \tag{25}$$

where D_{BL}, D_{BG} = diffusivity of B through the liquid and gas, respectively and δ_{mL}, δ_{mG} = thickness of a hypothetical layer of fluid on each side of the interphase.

These equations are the mass transfer analogues of equation 7 and many of the comments regarding heat transfer coefficients are applicable to h_{mL} and h_{mG}. This similarity has been used to estimate heat transfer coefficients from the results of mass transfer tests and vice versa. The analogy between the two transfer processes is evident when presenting some of the useful correlations of mass transfer coefficients. Dimensional analysis indicates that for mass transfer between a solid and a fluid:

$$Sh = f(Re, Sc, Gr) \tag{26}$$

where

$$Sh = \text{Sherwood number} = \frac{h_m L}{D}$$

$$Sc = \text{Schmidt number} = \frac{\mu}{\rho D}$$

and

$$Gr = \text{Grashof number} = \frac{g\,\Delta\rho L^3\rho^2}{\rho\mu^2}$$

Comparison with equation 8 suggests that the Sherwood and Schmidt numbers are analogous to the Nusselt and Prandtl numbers. Indeed experimentation has shown that the correlating functions are exactly, or very nearly, the same. For example, the transfer coefficients for turbulent flow along a pipe are correlated by:

$$Nu = 0.023\,Re^{0.8}Pr^{0.4} \text{ for heating} \tag{27}$$

$$Sh = 0.023\,Re^{0.83}Sc^{0.33} \text{ for mass transfer to liquids}^{53} \tag{28}$$

$$Sh = 0.023\,Re^{0.83}Sc^{0.44} \text{ for mass transfer to gases}^{53} \tag{29}$$

In using the heat/mass transfer analogy to estimate mass transfer coefficients, it must be ensured that analogous conditions exist under both circumstances, for example, the driving forces, ΔT and Δc, must act in the same direction and the geometry and flow regime, laminar or turbulent, must be identical. Due to the similarities of the correlations of heat and mass transfer coefficients at solid/fluid boundaries, no further consideration of them is required.

The simple concept of layers of fluid on each side of a fluid/fluid interface is obviously somewhat naive and more refined theories have been proposed for estimating mass transfer coefficients under these conditions.[54] These theories are useful for indicating the effect on the mass transfer coefficient of changing the operating conditions but the accuracy of the estimated coefficients is usually inadequate for design purposes and experimentation is necessary. Since it is impossible to measure the concentration of a solute at the interface, an overall mass transfer coefficient is defined in a similar way to the overall heat transfer coefficient, so that for a liquid/gas interface:

$$N_B = h_{mOL}A\,\Delta c_B \tag{30}$$

$$N_B = h_{mOG}A\,\Delta p_B \tag{31}$$

where h_{mOL}, h_{mOG} = overall mass transfer coefficients based on the effective concentration difference, or effective partial pressure difference, between the phases and A = total interfacial surface area. The significance of these overall coefficients is shown by illustrating their relationship to the individual coefficients. In the liquid and gas phases, the most easily measured driving force is used in defining the transfer coefficients, equations 22 and 23. In relating these driving forces to the effective overall driving forces, it is necessary to know the equilibrium relationship between the concentration of the solute in the liquid and its partial pressure in the gas, that is:

$$c_B^* = f(p_B) \tag{32}$$

or

$$p_B^* = f(c_B) \tag{33}$$

where c_B^*, p_B^* = concentration or partial pressure of solute that would exist in equilibrium with p_B or c_B.

These relationships will be considered later in more detail, but with the initial assumption:

$$p_B^* = mc_B \tag{34}$$

and substituting into equations 22 and 23, equations 30 and 31 apply such that:

$$\frac{1}{h_{mOL}} = \frac{1}{h_{mL}} + \frac{1}{mh_{mG}} \tag{35}$$

$$\frac{1}{h_{mOG}} = \frac{m}{h_{mL}} + \frac{1}{h_{mG}} \tag{36}$$

These equations show that the overall resistance to mass transfer, $(1/h_{mOL})$ or $(1/h_{mOG})$, is equal to the sum of the individual resistances and by producing a negligible resistance in one phase, for example by using a high gas velocity, it is possible to determine the transfer coefficient for the other phase. However, in using these individual coefficients to estimate h_{mOL} or h_{mOG} in other circumstances it must be ensured that all of the resistances are considered and that these resistances are independent. These requirements may be difficult to satisfy, for example, in liquid/liquid extraction when a solute transfers between two insoluble liquids, the liquid flows may affect each other and the use of individual coefficients obtained under other circumstances cannot be justified.

Equations 22, 23, 30 and 31 illustrate the need to provide large surface areas for mass transfer. This may be achieved with various processing equipment, such as spray, tray and packed towers and static and jet mixers. The choice of equipment is based on the properties of the materials, throughput rates, required degree of contacting, energy consumption and space limitations. In all these examples it is often difficult or impossible to measure the interfacial area and common design procedures use $h_{mOL}A$ or $h_{mOG}A$. Such values are specific to the geometry of the packing, the type of mixer, and the materials and flowrates.[42]

Many of the factors considered above are also relevant to the design of fermentation systems where air is passed into the fermentation broth. One of the main design criteria is to provide adequate oxygen transfer from the air bubbles to the microorganisms with minimum power requirements on the

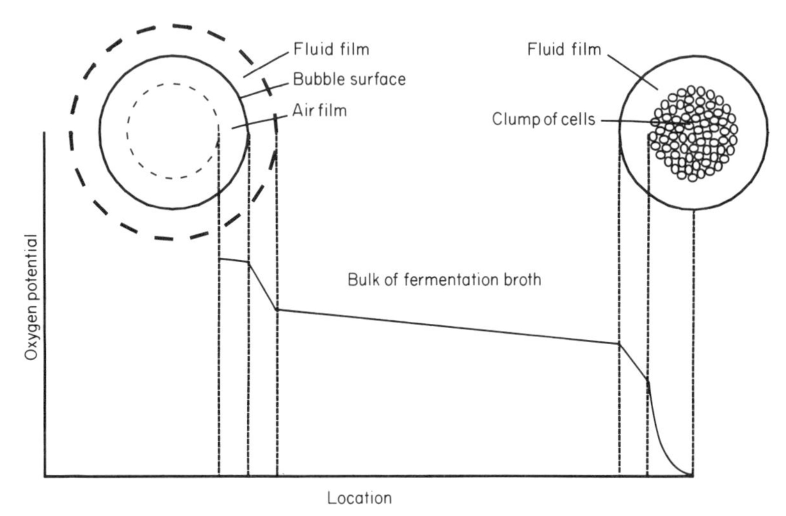

Fig. 3 Schematic diagram of the oxygen profile in a fermenter illustrating the series resistances

paddles used for mixing and splitting the bubbles. Figure 3 shows schematically the oxygen profile between a bubble and a microorganism. Examination of each transfer resistance[4] shows that for large cell flocs and no surfactant additives in the broth, the largest resistance to oxygen transfer lies in the liquid layers around the bubbles and the flocs, within the flocs and possibly within the bulk broth if mixing is insufficiently vigorous. Alternatively, if the cells are disperse, the fluid film resistances around the bubbles will be significant but there is a greater likelihood of cells attaching to bubbles and eliminating the bulk broth resistance. In addition, the larger surface areas of the fluid films around the disperse cells reduce the film resistance and the diffusion path within the cells is also reduced. Although the main resistances have been considered it is important to recognize the others, namely, that within the air film inside the bubble and possible resistance caused by additives, such as antifoaming agents, attaching to the surface of the bubbles.

The rates of many food processing operations are controlled by factors similar to those above. It is always important to consider all of the transfer resistances and then design the processing equipment to reduce those which are dominant, for example, by mixing, using additives or varying the process conditions such as temperature or pressure.

C Phase equilibria and phase changes

The previous discussion has shown that mass transfer occurs within a single phase when a concentration gradient exists within that phase. If the transfer process is allowed to continue indefinitely, the concentration within the phase will become uniform. Similarly, in a multiphase system the ultimate concentration gradient within each phase will be zero but the concentration of solute will be different in each phase. The relationship between these final concentrations is the equilibrium relationship or distribution and this represents the greatest possible degree of separation. Equations 32, 33 and 34 are equilibrium relationships of a solute, B, between a gas and a liquid. These functions were used in relating the individual and overall mass transfer coefficients for a particular system which was not at phase equilibrium, i.e. concentration gradients did exist. However, the relationships were valid at the liquid/gas interface where it was assumed there was no transfer resistance and therefore a phase equilibrium prevailed.

Equilibrium relationships are usually determined experimentally or, in cases when specific data are not available, the laws of thermodynamics are used to obtain estimations often based on experimental data measured with similar systems. Discussion of these techniques is beyond the scope of this presentation but details may be found in many texts of thermodynamics[5] and useful bibliographies of equilibration data for chemical systems include Wichterle *et al.*[56] and Francis.[18] In foodstuffs the transfer of water is particularly important because moisture losses affect the value of the food and many operations which increase shelf life do so by reducing the availability of water to microorganisms. For this reason, the remainder of this section is concerned with the equilibrium relationships of water with emphasis on their relevance to foodstuffs.

Consider a sample of chilled food placed in a warm air environment. If the surface temperature of the food is below the dew point of the air, water condenses on the surface and the latent heat evolved increases the surface temperature. The equilibrium relationships of pure water apply at the vapour/liquid interface and the effective partial pressure difference of water between these phases causes the water to evaporate and later in the warming process it is the water from the food that is lost to the air. At this time, the pertinent equilibrium relationship at the interface is different from that of pure water because the moisture in foods contains dissolved substances that depress the vapour pressure of the water. This depression increases the boiling point and decreases the freezing point of water in foods. Also, the vapour pressure may be depressed because the water exists in small capillaries. Some of the water, often called "bound" water, is particularly difficult to remove from the food and this also exerts a reduced vapour pressure. In addition, some water is not "strongly bound" nor available for dissolving some of the food components. It is the difficulty of removing the "bound" water that leads to the large changes in the effective diffusion coefficients during drying.[22] The locations and interactions of the water in foods are considered by Duckworth[17] and Rizvi and Benado.[46] The complex relationships between the water in the food and the surrounding environment are usefully described by water sorption isotherms—plots of moisture content versus water activity (partial pressure/saturated vapour pressure). Various mathematical relationships have been used to correlate the isotherms over restricted ranges of water content for different temperatures and foods[26,46] and these are often used in models of mass transfer.

The need to add or remove heat during phase changes is fundamental to most food processing operations. For example, combined heat and mass transfer occurs during freezing, thawing, freeze drying and drying. In most operations, the difficulties of mass transfer are insignificant compared to those of transferring large amounts of heat required for the phase change, for example, the crystallization of water during the freezing of foods. However, with other processes such as drying, the transfer of water away from the phase change zone may restrict the rate of processing because of the difficulty in moving water through the food. To illustrate these important phenomena and bring together much of the material presented earlier, a few examples of processing operations will be considered briefly.

Food may be frozen by passing cold air across its surface. Heat is transferred outwards through the food by conduction and away from the surface by convection and, to a far less an extent, by radiation. After the food temperature has fallen below the freezing point of the water within the food, small groups of ice crystals may form and these crystals grow as freezing proceeds. During these early stages, water can move relatively freely through the solids network and the freezing process is limited by the rate of removal of the latent heat of crystallization and sensible heat. However, with reducing temperatures and increasing concentrations of the solutes in the water, the movement of the fewer remaining water molecules is more restricted and may control the rate of freezing. It is difficult to increase the rate of this mass transfer and fortunately it is unnecessary in most commercial operations because the food may be considered as frozen before the mass transfer is restricted. Consequently freezing is usually considered as a heat transfer process.

In drying, heat and mass transfer need to be considered simultaneously. Often heat is transferred to the food by convection, radiation and a little by conduction along any supports. The relatively low thermal conductivity of most foods restricts the rate of heat conduction into the food and consequently large amounts of heat are available for increasing the surface temperature and for evaporation. The evaporation zone will move into the food if the rate of internal moisture diffusion is insufficient to replace the loss by evaporation. Regions of liquid and vapour diffusion then exist within the food and analysis of the drying operation needs to include descriptions of these flows as well as the boundary conditions and the transfer of heat.

Cooking in air may be considered in the same manner as dehydration and in both cases it is common for the food to shrink, complicating any mathematical analysis, and more importantly it may reduce the quality of the food. Shrinkage during dehydration can be reduced by freezing the food and then subliming the ice (freeze drying). Essentially, this process is described by the heat and mass transfer analysis relating to drying but without internal liquid diffusion and with sublimation replacing evaporation. As with many operations, there is a possible loss of quality and the cost of processing may be prohibitive.

4 CONCLUSIONS

The equations presented are models used to represent heat and mass transfer and to enable ideas to be described and utilized. In some cases, for example with the concept of a boundary layer and a surface transfer coefficient, the models are obviously simplications of the real events. The similarity between the descriptions of heat and mass flows and the consequent use of identical mathematical techniques to solve both sets of equations have been noted. Most of these equations and solutions were initially developed to describe the transfer of heat through common engineering materials. However, this early work has been applied, with varying success, to the analysis of the many food processing operations. The extension into foods has introduced further complications such as shrinkage and rapidly varying properties which depend on previous treatments. Data on thermophysical properties and surface transfer coefficients need to be determined for many operations. A food technologist is concerned not only with processing rates and energy costs but with transferring heat and mass to increase storage life and retain and improve quality characteristics such as juiciness, tenderness, nutritional value and flavour. The concepts underlying the heat and mass transfer phenomena and their effect on the characteristics of foods need to be understood before existing processes can be improved and new more efficient, and perhaps novel, operations can be designed.

Nomenclature

A	cross sectional or surface area
a	geometric factor in equations 2 and 20: for a slab, $a = 0$; for a cylinder, $a = 1$; for a sphere, $a = 2$
Bi	Biot number $= hX/k$
c	specific heat capacity
c_B	concentration of substance B
d, d_0	inside and outside diameter

d_{LM}	logarithmic mean diameter $= (d_0 - d)/\ln(d_0/d)$
D	mass diffusion coefficient
E	field strength, V^{-1} cm
F_A, F_ε	geometry and emissivity factor in equation 29
g	acceleration due to gravity
Gr	Grashof number; for heat transfer, $Gr = \beta g\, \Delta T L^3 \rho^2/\mu^2$; for mass transfer, $Gr = g\, \Delta\rho L^3 \rho^2/\rho\mu^2$
h	heat transfer coefficient
h_m	mass transfer coefficient
k	thermal conductivity
l	length
L	characteristic dimension
m	equilibrium distribution constant
N	rate of mass transfer
Nu	Nusselt number $= h_c L/k$
p	partial pressure
P	power, W
Pr	Prandtl number $= c\mu/k$
q	rate of heat transfer
Re	Reynolds number $= Lu\rho/\mu$
Ri, Ro	scale resistances to heat transfer on the inside and outside surfaces of a pipe
Sc	Schmidt number $= \mu/\rho D$
Sh	Sherwood number $= h_m L/k$
t	time
T	temperature
T_1, T_2	surface temperature of objects 1 and 2
u	fluid velocity
U	overall heat transfer coefficient
V	volume of material
x	distance
x_{Ri}, x_{Ro}	thickness of scale layers on inside and outside surface of the pipe
X	radius or half thickness
β	coefficient of thermal expansion
γ	parameter defined by equation 11
δ	thickness of boundary layer
δ_{mL}, δ_{mG}	thickness of the mass transfer boundary layer in a liquid or gas
Δc	concentration difference
Δp	partial pressure difference
ΔT	temperature difference
ΔT_{LM}	logarithmic mean temperature difference
$\Delta\rho$	fluid density difference between the surface and fluid bulk

ε_r	effective loss factor
μ	viscosity
μ, μ_w	viscosity evaluated at the mixed mean fluid temperature or the solid surface temperature, respectively
ρ	density
σ	Stefan-Boltzmann constant $= 5.68 \times 10^{-8}$ W m^{-2}K^{-4}
ω	electrical frequency

SUBSCRIPTS

B	substance B
c	convection
$cond$	conduction
r	radiation
G	gas
L	liquid
Ri, Ro	fluid/scale interface inside and outside of the pipe
w	wall

References

1. Abdul Majeed, P. M., Srinivasa Murthy, S. and Krishna Murthy, M. V. (1980). *Trans. ASAE* **23**, 788–792.
2. Adam, M. (1969). "Bibliography of Physical Properties of Foodstuffs", C.A.Z., Prague, Czechoslovakia.
2. Baghe-Khandan, M. S., Okos, M. R. and Sweat, V. E. (1982). *Trans. ASAE* **25**, 1118–1122.
4. Bailey, J. E. and Ollis, D. F. (1977). Biochemical Engineering Fundamentals, pp. 413–418, McGraw-Hill, New York.
5. Bett, K. E., Rowlinson, J. S. and Saville, G. (1975). Thermodynamics for Chemical Engineers, MIT Press, Cambridge, Massachusetts.
6. Bird, R. B., Stewart, W. B. and Lightfoot, E. N. (1960). Transport Phenomena, pp. 571, John Wiley and Sons, New York.
7. Burfoot, D. (1982). The turbulent single phase forced convection heat transfer and pressure drop characteristics of circular ducts containing swirl flow inducers or Pall rings, 15–49, PhD Thesis, Loughborough University of Technology.
8. Burfoot, D. (1983). Problems in mathematically modelling the cooking of a joint of meat, Proc. COST 91, Athens (in press).
9. Cleland, A. C. and Earle, R. L. (1979). *J. Fd Sci.* **44**, 958–963.
10. Cleland, A. C. and Earle, R. L. (1979). *J. Fd Sci.* **44**, 964–970.
11. Cleland, D. J., Cleland, A. C., Earle, R. L. and Byrne, S. J. (1984). *Int. J. Refrig.* **7**, 6–13.
12. Coulson, J. M. and Richardson, J. F. (1970). Chemical Engineering 1, 2nd edn, pp. 345–392, Pergamon Press, Oxford.
13. Coulson, J. M. and Richardson, J. F. (1970). *ibid*, 199–201.
14. Dagerskog, M. (1979). *Lebensm.-Wiss. u.-Technol.* **12**, 217–224.

15. Dagerskog, M. (1979). *Lebensm.-Wiss. u.-Technol.* **12**, 225–230.
16. Dalgleish, N. and Ede, A. J. (1965). Charts for determining centre, surface, and mean temperatures in regular geometric solids during heating or cooling, National Engineering Laboratory Report No. 192, Glasgow, Scotland.
17. Duckworth, R. B. (1975). Water relations of foods, Academic Press, London.
18. Francis, A. W. (1963). Liquid-Liquid Equilibriums, John Wiley/Interscience, New York.
19. Heldman, D. R. and Singh, R. P. (1981). Food Process Engineering, 2nd edn, pp. 124–142, AVI Publ., Westport, Connecticut.
20. Housova, J. (1977). The baking process—mathematical model of heat and mass transfer, Proc. EFCE Mini-symposium in Food Processing, Orenas, Sweden, 249–267.
21. James, S. J., Creed, P. G. and Roberts, T. A. (1977). *J. Sci. Fd Agric.* **28**, 1109–1119.
22. Jason, A. C. (1958). A study of evaporation and diffusion processes in the drying of fish muscle, Session IV of SCI Conference on Fundamental Aspects of the Dehydration of Foodstuffs, Society of the Chemical Industry, London.
23. Jason, A. C. and Long, R. A. K. (1955). The specific heat and thermal conductivity of fish muscle, Proc. 9th Int. Congr. Refr., Paris, 2.160–2.169.
24. Johns, P. B. (1987). Modelling heat and mass transfer in foodstuffs, in Food structure and behaviour: Equilibrium and non-equilibrium aspects, Blanshard, J. M. V. and Lillford, P. J., Royal Society of Chemistry, London.
25. Jowitt, R., Escher, F., Hallstrom, B., Meffert, H. F. T., Spiess, W. E. L. and Vos, G. (1983). Physical properties of foods, Applied Science, London.
26. Karel, M. (1973). *CRC Crit. Rev. Food Technology* **3**, 329–373.
27. Kays, W. M. (1966). Convective Heat and Mass Transfer, p. 196, McGraw-Hill, New York.
28. Kern, D. Q. (1950). Process Heat Transfer, pp. 639–664, McGraw-Hill, Kogakusha, Tokyo.
29. Kern, D. Q. (1950). *ibid.* 87–97.
30. Kern, D. Q. (1950). *ibid.* 840.
31. Kern, D. Q. (1950). *ibid.* 78–82.
32. Knudsen, J. G. and Katz, D. L. (1958). Fluid Dynamics and Heat Transfer, p. 403, McGraw-Hill, New York.
33. Krishna Murthy, M. V. and Badari Narayana, K. (1978). Heat and mass transfer in food products, 6th Int. Heat Transfer Conf., Toronto, **6**, 339–354, Hemisphere Publ.
34. Liapis, A. I. and Litchfield, R. J. (1979). *Computers and Chemical Engineering* **3**, 615–621.
35. McAdams, W. H. (1933). Heat Transmission, 1st edn, pp. 26–39, McGraw-Hill, New York.
36. Miles, C. A. (1982). Heat transfer at the air/meat interface, Meat Research Institute Memorandum No. 53, Langford, Bristol, England.
37. Miles, C. A., van Beek, G. and Veerkamp, C. H. (1983). Physical Properties of Foods, eds Jowitt, R., Escher, F., Hallstrom, B., Meffert, H. F. Th., Spiess, W. E. L. and Vos, G., pp. 269–312, Applied Science, London.
38. Morley, M. J. (1972). Thermal Properties of Meat: Tabulated Data, Meat Research Institute Special Report No. 1, Langford, Bristol, England.
39. Perry, R. H. and Chilton, C. H. (1973). Chemical Engineers' Handbook, 5th edn, pp. 10–11, McGraw-Hill Kogakusha, Tokyo.
40. Perry, R. H. and Chilton, C. H. (1973). *ibid.* 10–15.

41. Perry, R. H. and Chilton, C. H. (1973). *ibid.* 3_230–3_235.
42. Perry, R. H. and Chilton, C. H. (1973). *ibid.* 18_19–18_49.
43. Prentice, J. H. and Huber, D. (1983). Physical Properties of Foods, eds Jowitt, R., Escher, F., Hallstrom, B., Meffert, H. F. Th., Spiess, W. E. L. and Vos, G., pp. 123–183, Applied Science, London.
44. Ramaswamy, H. S., Lo, K. V. and Tung, M. A. (1982). *J. Fd Sci.* **47**, 2042–2047, 2065.
45. Riedel, L. (1956). *Kaltetchnik* **8**, 374–377.
46. Rizvi, S. S. H. and Benado, A. L. (1984). *Fd Techn.* **38**, 83–92.
47. Schlichting, H. (1979). Boundary Layer Theory, 7th edn, McGraw-Hill, New York.
48. Schweiger, R. G. (1974). Meat and meat products, IFST Mini-symposium, 10–14, Institution of Food Science and Technology, London.
49. Skelland, A. H. P. (1967). Non-Newtonian Flow and Heat Transfer, pp. 358–433, John Wiley and Sons, New York.
50. Smith, R. B. (1974). *Trans. IMPI* **2**, 1–113.
51. Smith, R. B. (1975). Microwave Heating for the Food Industry, Short Course at University of Bradford, pp. 1–68, IMPI-Europe, London.
52. Sorenfors, P. (1977). Heat and mass transfer in frying, Proc. of EFCE Mini-symposium in Food Processing, Orenas, Sweden, 289–300.
53. Treybal, R. E. (1968). Mass Transfer Operations, 2nd edn, p. 62, McGraw-Hill, Kogakusha, Tokyo.
54. Treybal, R. E. (1968). *ibid.* 49–54.
55. Uhl, V. W. (1970). Augmentation of Convective Heat and Mass Transfer, eds Bergles, A. E. and Webb, R. L., pp. 109–117, American Society of Mechanical Engineers, New York.
56. Wichterle, I., Linek, J. and Hala, E. (1973). Vapour-Liquid Equilibrium Data Bibliography, Elsevier, Amsterdam.
57. Yamada, T. (1970). *Agr. Chem. Soc. Japan J.* **44**, 587–590.

12 Flavour Release—Elusive and Dynamic

P. B. McNULTY

Agricultural and Food Engineering Department, University College, Dublin

1 INTRODUCTION

The perceived flavour of foods may be significantly affected by differences in the rates and extent of flavour release in the mouth. This problem has been investigated both theoretically and experimentally as reported in a series of papers by McNulty and co-workers. The objective of this study is to review these papers with particular emphasis on the effect of flavour release on the sensory perception of food. Included in the discussion will be the effect of physical properties, structure and processing of foods on their flavour-release characteristics.

The importance of flavour release in sensory characterization of foods has been recently discussed by Forss.[4] Lillford[7] has indicated the importance of flavour release in the development of new food products.

FOOD STRUCTURE AND BEHAVIOUR
ISBN 0-12-104230-8

2 DESCRIPTION OF RELEASE MODEL

McNulty and Karel[13] have developed a flavour release model which simulates flavour release in the mouth from o/w emulsions. The model assumes that: (a) flavour compounds are transferred from oil to water when the interphase equilibria are disturbed by dilution with saliva; and (b) only the aqueous concentration stimulates perception (Fig. 1). Flavour distribution and release equations were derived as follows:

$$C_{we} = \frac{C_{owi}}{\phi_i(K_p - 1) + DF_{em}} \tag{1}$$

$$C_{wd} = \frac{C_{wei}(1 - \phi_i)}{DF_{em} - \phi_i} \tag{2}$$

where C_{we} = equilibrium aqueous flavour concentration; C_{owi} = initial emulsion flavour concentration: ϕ_i = initial oil volume fraction; DF_{em} = emulsion dilution factor; C_{wd} = aqueous flavour concentration in the mouth immediately after emulsion dilution with saliva; C_{wei} = equilibrium aqueous concentration prior to dilution and the equilibrium partition coefficient, K_p is given in equation 3:

$$K_p = C_{oe}/C_{we} \tag{3}$$

where C_{oe}, C_{we} = flavour concentrations in the oil and aqueous phases at equilibrium respectively. When $DF_{em} = 1$ (i.e. zero dilution) equation 1 gives the initial equilibrium aqueous flavour concentration C_{wei}. When $DF_{em} > 1$ equation 1 gives the potential equilibrium aqueous concentration in the mouth which may be attained if equilibrium is restored.

The potential extent of flavour release has been defined as the dimension-

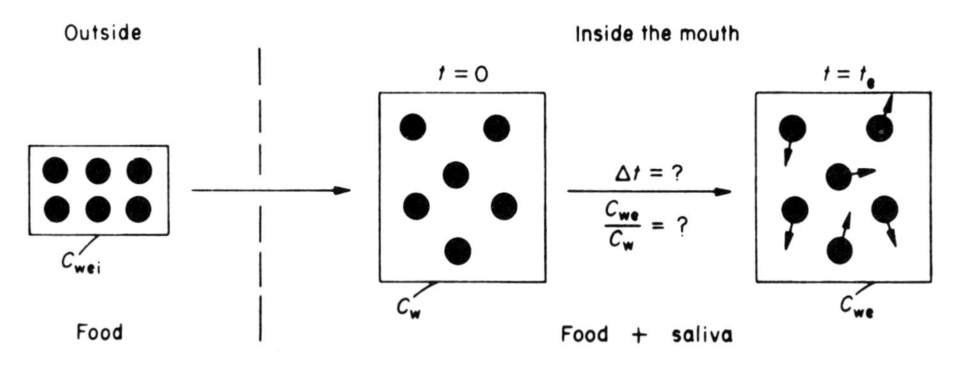

Fig. 1 Simulation of oral consumption of a simple model o/w emulsion food. If aqueous flavour only is perceived, we can ask: (a) how much flavour can be released?; and (b) how fast is the rate of transfer?

less parameter, C_{we}/C_{wd}, which is given by:

$$\frac{C_{we}}{C_{wd}} = \frac{[\phi_i(K_p - 1) + 1](DF_{em} - \phi_i)}{[\phi_i(K_p - 1) + DF_{em}](1 - \phi_i)} \tag{4}$$

The potential extent of release is also given by:

$$\frac{C_{we}}{C_{wd}} = \frac{\bar{C}_{we}}{\bar{C}_{wd}} = \frac{\bar{C}_{we}}{\bar{C}_{wei}} \tag{5}$$

where $\bar{C}$ gives total concentration in μg and C gives flavour concentration per unit volume, ppm or μg/ml. The total equilibrium aqueous flavour concentration, C_{we}, is given by:

$$\bar{C}_{we} = C_{we}V_{owi}(DF_{em} - \phi_i) \tag{6}$$

where V_{owi} = initial emulsion volume. The total flavour which is released from oil to water if equilibration is restored following aqueous dilution is given by $\Delta\bar{C}_w$ where:

$$\Delta\bar{C}_w = \bar{C}_{we} - \bar{C}_{wei} \tag{7}$$

3 MILK AND MAYONNAISE

Let us now apply the release equations 1–7 to two commercial food emulsions with extreme values of ϕ_i, i.e. milk ($\phi_i = 0.035$) and mayonnaise ($\phi_i = 0.80$). If we assume that $C_{owi} = 100$ ppm, $DF_{em} = 2$ (i.e. a 1:1 dilution) and $V_{owi} = 10$ ml, then we can evaluate flavour concentrations in the aqueous phases of milk and mayonnaise after dilution with saliva. Table 1 reveals that as K_p increases the potential extent of flavour release increases slightly for milk and more appreciably for mayonnaise (see also Fig. 2). Of greater significance, however, is the effect of K_p on the initial inter-phase flavour

Table 1 Effect of the equilibrium partition coefficient, K_p on potential flavour release after emulsion dilution ($DF_{em} = 2$; $C_{owi} = 100$ ppm; $V_{owi} = 10$ ml)

K_p	milk ($\phi_i = 0.035$)				mayonnaise ($\phi_i = 0.8$)			
	C_{wd} (ppm)	$\frac{C_{we}}{C_{wd}}$	$\%\frac{\bar{C}_{wei}}{\bar{C}_{ow}}$	$\%\frac{\Delta\bar{C}_w}{\bar{C}_{ow}}$	C_{wd} (ppm)	$\frac{C_{we}}{C_{wd}}$	$\%\frac{\bar{C}_{wei}}{\bar{C}_{ow}}$	$\%\frac{\Delta\bar{C}_w}{\bar{C}_{ow}}$
10^{-1}	50.5	1.01	98.8	0.99	59.7	1.31	71.5	22.2
10^0	49.0	1.02	96.3	1.93	16.7	3.00	20.0	40.0
10^1	37.2	1.16	73.2	11.7	2.04	5.34	2.45	10.6
10^2	11.0	1.66	21.7	14.3	0.21	5.91	0.25	1.23
10^3	1.4	1.98	2.7	2.64	0.02	5.99	0.02	0.12

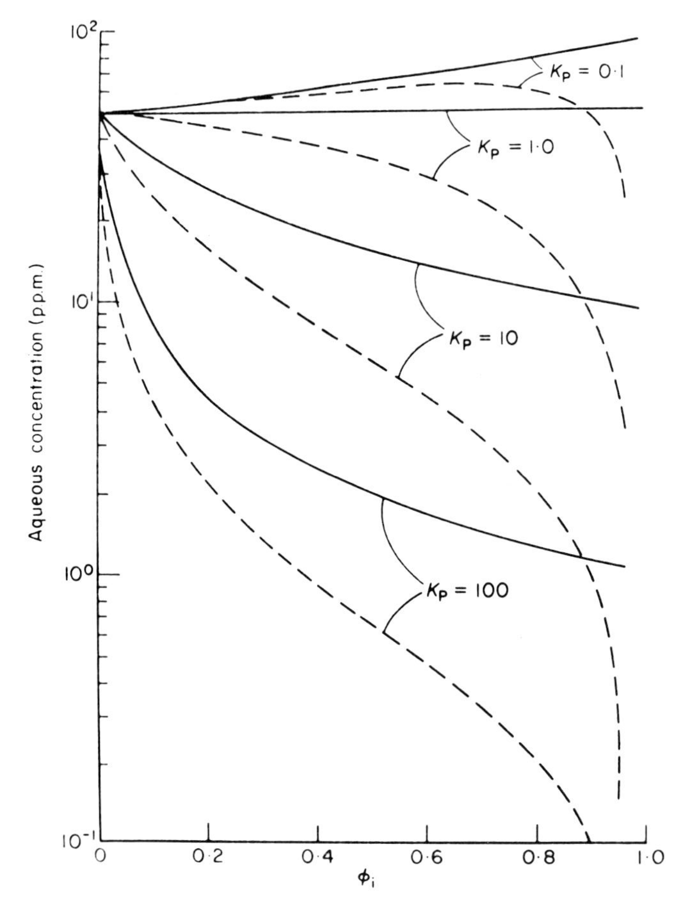

Fig. 2 Effect of K_p and ϕ_i on the aqueous concentration immediately after emulsion dilution, C_{wd}, and at equilibrium, C_{we}, for $DF_{em} = 2$ and $C_{owi} = 100$ ppm

distribution as indicated by the parameter $\%\bar{C}_{wei}/\bar{C}_{ow}$. These data reveal that at low K_p the flavour is distributed primarily in the aqueous phase (especially in the case of milk) and therefore flavour release is relatively unimportant. Flavour release becomes important as K_p increases because: (a) the potential extent of flavour release increases; and (b) the distribution of flavour in the aqueous phase decreases. Flavour release becomes important at lower values of K_p in mayonnaise than in milk.

In general, we may conclude that: (a) flavour distribution or "fixation" in oils and fats increases as both ϕ_i and K_p increase; and (b) the potential extent of flavour release in the mouth increases as K_p, ϕ_i and/or emulsion dilution increase.[10,13]

4 EXPERIMENTAL STUDIES

Interphase distribution and the extents and rates of release of model flavours were measured in both stirred diffusion cells[14] and in o/w emulsions following a step-change aqueous dilution.[15] The model flavour compounds were *n*-alcohols (C_3–C_8) and cholesterol.

A Flavour distribution and extent of release

The equilibrium partition coefficient determines both inter-phase flavour distribution and the extent of release for constant values of ϕ_i and DF_{em}. The various factors causing K_p to vary in oil-water systems in stirred diffusion cell experiments are discussed below. Some experimental K_p values are given in Tables 2 and 3.

Table 2 Effect of the solid fat index (SFI) of vegetable oils and fats on transfer rates of *n*-hexanol at 24 °C in a stirred diffusion cell ($\phi_i = 0.5$)[a]

SFI	no surfactant		+aqueous Tween 60	
	K_p	rate × 10^4 (s^{-1})	K_p	rate × 10^4 (s^{-1})
0	9.00	0.539	7.33	0.464
8	8.09	0.225	6.69	0.188
24	9.53	0.263	8.09	0.158
66	6.14	0.0283	4.88	0.0503

[a] Data abstracted from McNulty[9].

Table 3 Effect of oil viscosity, η_0, on transfer rates of *n*-octanol at 24 °C in a stirred diffusion cell ($\phi_i = 0.5$)[a]

oil			no surfactant		+aq Tween 60	
type	η_0 (cp)	SFI	K_p	rate × 10^4 (s^{-1})	K_p	rate × 10^4 (s^{-1})
vegetable	62	0	99.0	1.18	50.0	0.633
silicone[b]	480	0	26.0	1.24	10.1	0.728
silicone[b]	4 750	0	31.5	0.559	9.0	0.107
silicone[b]	93 000	0	25.4	0.635	9.0	0.102
tristearin	∞	100	44.4	0.032	24.0	0.019

[a] Data abstracted from McNulty[9].
[b] Linear polydimethylsiloxanes.

Effect of the oil solid fat index (SFI). In the case of *n*-hexanol, K_p decreased slightly with increase in SFI in the presence and absence of surfactant (Table 2). A more pronounced decrease in K_p occurred when *n*-octanol K_p values for liquid vegetable (SFI = 0) and tristearin (SFI = 100) oils are compared (Table 3).

Effect of oil type and viscosity. In silicone oils ($48 < \eta_0 < 93000$ cp), oil viscosity appeared to have little effect on the K_p of *n*-octanol both in the presence and absence of surfactant. All K_p values were appreciably lower than those for vegetable oil (Table 3).

Effect of salt. The addition of 5% sodium sulphate to the aqueous phase, increased the K_p of *n*-octanol between silicone oil and water from 26.8–46.6, whereas the addition of surfactant to the same system decreased K_p from 26.8–9.53.[14]

Effect of aqueous surfactant. In all cases the presence of aqueous Tween 60 reduced K_p values (Tables 2 and 3). The effect was more pronounced for octanol than hexanol. McNulty[9] has suggested that aqueous alcohol/surfactant associations result in the reduction of K_p values and that such associations increase in magnitude as the alcohol chain length increases, i.e. as aqueous solubility of the alcohol decreases.

B Rates of flavour release

Effect of the oil solid fat index (SFI). Release rates of *n*-hexanol decrease as SFI increased in the presence and absence of surfactant (Table 2). The relatively small depression in flavour release from semi-solid oils is probably due to flavour diffusion occurring primarily in the liquid portion of the oil. In the case of tristearin (SFI = 100, Table 3), solidification of the oil is accompanied by shrinkage which probably results in an amorphous and therefore permeable structure.

Effect of oil viscosity, η_0. Release rates of *n*-octanol decreased only slightly as η_0 was increased (Table 3) in contradiction to a prediction based on diffusion theory.[14] All results are for silicone oils except those for vegetable oil ($\eta_0 = 62$ cp) and tristearin ($\eta_0 = \infty$). The apparent permeability of silicone oils to *n*-octanol is probably due to their rather unique chemical structure which has been discussed by McNulty and Karel.[14]

C Emulsion studies

Kinetic studies[15] revealed that flavour release rates are normally rapid in non-aggregated liquid o/w emulsions stabilized by mixed surfactant systems, i.e. equilibration half-lives are normally less than 15 s. These extremely rapid release rates are primarily due to the enormous interfacial area available for mass transfer in o/w emulsions. Therefore, if high values of K_p and/or ϕ_i exist in the mouth the total flavour will be: (a) partitioned primarily in the oil phase; and (b) released rapidly from the oil droplets following dilution with saliva, and thus become available for perception. This conclusion was qualitatively supported by the rapid sensory perception of such primarily oil-soluble compounds as ethanol and benzaldehyde (2–5 s), and caffeine and menthol (5–15 s) found when o/w emulsions ($\phi_i = 0.1$) were tasted in the mouth.[9] The complicating effect of aqueous flavour-surfactant associations on sensory perception has been demonstrated by McNulty and Moscowitz.[16] Solute-surfactant associations were also found to be responsible for the extremely slow interphase transport of cholesterol under certain experimental conditions.[11,15]

Flavour release in the mouth can be greatly retarded by increasing ϕ_i until the emulsion is "plastic" (e.g. mayonnaise), because mixing with saliva will probably be slow. In these cases, flavour release rate will depend not only on the physical-chemical emulsion properties but also on the bite size, the extent of mastication and mixing, and the residence time in the mouth. Similar considerations apply to foods with semi-solid and solid structures.

5 FLAVOUR THRESHOLD DATA

Forss[4] has reviewed the threshold data literature for odour and flavour compounds from lipids in various media and has repeated the well known observation that odour and flavour potentials are normally much stronger in aqueous than in oily media. He also suggested some qualitative theories by which this phenomenon can be explained but no quantitative theory was proposed.

McNulty and Karel[13] have applied the release model to the flavour threshold data for *n*-alkanals reported by Lea and Swoboda[5] (Fig. 3). The following assumptions were used in the analysis: (a) each flavoured medium mixes on a 1:1 basis with saliva in the mouth (i.e. $DF_{em} = 2$); (b) only the flavour in the aqueous phase can stimulate perception; (c) oil and water are extreme cases of o/w emulsions with values of ϕ_i of 1.0 and zero respectively; and (d) rapid flavour equilibration occurs between oil and saliva. The latter assumption is probably valid because Lea and Swoboda administered the oil samples into the mouth by means of a nasal atomizer, thus ensuring a large surface area for mass transfer.

By using equations 1–3 and the data in Fig. 3, it was possible to estimate the equilibrium oil-water partition coefficient, K_p, as described by McNulty.[9] The results were compared with some literature values [17] and good agreement was obtained at low chain lengths (C_3–C_6).

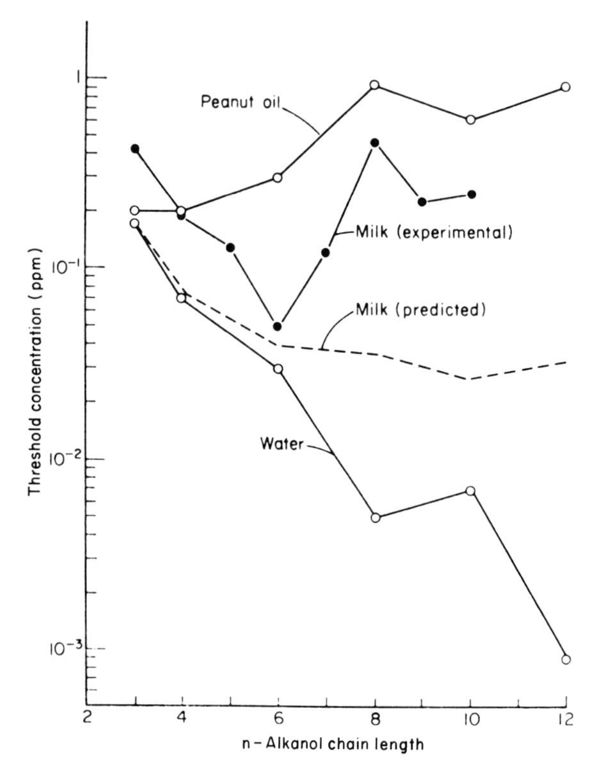

Fig. 3 Effect of solvent medium on threshold concentrations of *n*-alkanals

It was also possible to predict *n*-alkanal threshold concentrations in o/w emulsions such as milk by: (a) using the threshold data in oil and water; (b) using the values of K_p calculated therefrom; (3) assuming $DF_{em} = 2$ and $\phi_i = 0.035$ (in the case of milk); and (d) using the flavour release equations. The results compare reasonably well with the experimental values reported by Lillard *et al.*[6] particularly at low chain lengths (Fig. 3).

6 PERCEPTION OF ACIDS

Figure 4 presents flavour thresholds of *n*-alkanoic acids in various media. Threshold data in water from different laboratories[19,22] are in remarkably good agreement. Comparison of oil and water thresholds suggested that the lower chain acids, acetic (C_2), butyric (C_4) and hexanoic (C_6), had higher flavour potential in oil than in water thus reversing the general principle

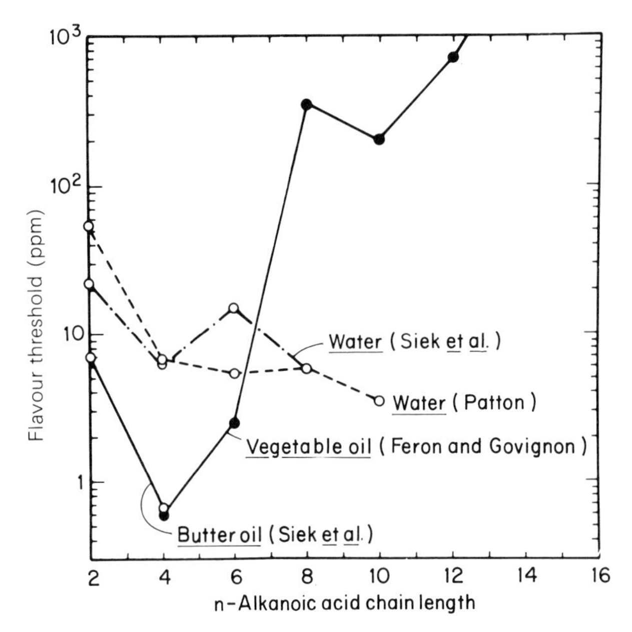

Fig. 4 Effect of solvent medium on threshold concentrations of *n*-alkanoic acids

previously stated. Hence the flavour release model could not be used to evaluate partition coefficients nor to predict threshold concentrations in o/w emulsions. Various proposals have been suggested to explain this unexpected phenomenon such as hydrogen bonding in aqueous solution,[4,19] salt formation in aqueous solution[4,8] and dimerization in aqueous solution.[4]

According to the flavour release model (which deals specifically with perception of taste in the mouth and thereby excludes odour perception), flavour compounds in oil are released into the aqueous saliva phase as a necessary precondition for perception. Therefore, such flavour compounds will be subjected to hydrogen bonding, salt formation and/or dimerization in just the same way as the flavours tasted from aqueous solution.

Thus the higher flavour potential of low-chain acids in oil over that in water is probably due to a higher odour potential and not to a higher taste potential. This conclusion is supported by odour thresholds for butyric (4.0 ppm) and hexanoic (8.8 ppm) acids[20] which are in close agreement with their aqueous flavour thresholds (Fig. 4). Nevertheless, this conclusion must be viewed with caution until such time as the flavour thresholds in oil and water are evaluated by one group of investigators to ensure that the comparison in Figure 4 is, in fact, valid.

Other flavour threshold data have been discussed by McNulty.[12]

7 EFFECT OF FOOD STRUCTURE ON SENSORY PERCEPTION

The release model was applied to the fascinating problem posed by Patton[19] as follows: "one of the most critical ways of evaluating butter fat, or any edible fat for that matter, is to appraise flavour after re-emulsification of the fat in good quality fluid skim-milk. This test appears far more sensitive than evaluation of the fat as such".

Figure 5 shows a proposed mechanism of flavour perception in butter and in butteroil re-emulsified in skim-milk. The top left hand compartment reveals an idealized butter structure in the form of a w/o (water-in-oil) emulsion. When this butter is sampled in the mouth it contacts but does not mix with saliva because its continuous phase is oil. Therefore, at zero time, the flavour concentration in the saliva is zero. As time progresses flavour is transferred from oil to saliva and at equilibrium the aqueous concentration will have increased from 0–1.23 ppm as shown in the top right-hand compartment. Stirred diffusion cell data[14] have revealed that the transport rate is extremely slow, e.g. the half-life for the flavour equilibration is approximately 2 h. From this figure the concentration after 15 s in the mouth was calculated to be approx. 0.01 ppm.

Let us now take a sample of butter and re-emulsify it in skim-milk as shown by the compartment in the lower left-hand corner. The proportions of butter and skim-milk have been arbitrarily chosen to give $\phi_i = 0.05$. Because

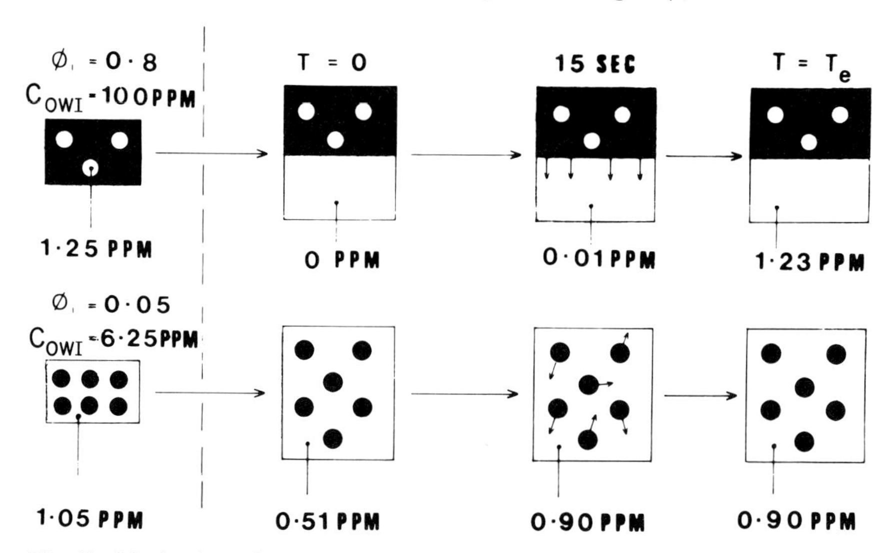

Fig. 5 Mechanism of perception of a flavour compound ($K_p = 100$) in butter and butteroil re-emulsified in skimmilk

the proportion of butter in the sample is small, C_{owi} is reduced from 100–6.25 ppm. When this o/w emulsion is sampled in the mouth it will immediately mix with saliva because its continuous phase is water. If the mixing is complete and instantaneous, the resulting aqueous concentration will be 0.51 ppm.

As time progresses more flavour will be transferred from oil to water. As previously stated the equilibration half-lives in such liquid o/w emulsion systems are normally less than 15 s. Thus we can assume that equilibration will probably have occurred after 15 s thereby increasing the aqueous concentrations from 0.51–0.90 ppm.

Figure 5 shows that, after 15 s, 0.90 ppm of flavour is available for perception in the butteroil-skimmilk system whereas only 0.01 ppm is available for perception in the original butter. Thus it is not surprising that flavour should be more easily perceived in butteroil re-emulsified in skimmilk than in butter itself.

This analysis was repeated for other values of partition coefficients and initial oil volume fractions. The absolute and relative magnitudes of the aqueous flavour concentrations varied considerably but the trend remained the same. In conclusion, we can now fully support with quantitative data the well-known observation that the "off-flavour" products of lipid oxidation are more easily perceived in o/w emulsions than in w/o emulsions or in the pure oil or fat phase itself.

8 DEHYDRATION PROCESSING

The physical processing of foods usually alters one or more of their physical properties and thereby may alter both flavour release and structural characteristics. In addition, certain processing operations may result in loss of desirable volatile flavour compounds. Thus in dehydration processing, for example, flavour retention is a desirable processing objective, whereas rapid flavour release is normally a desirable quality attribute during food consumption. This aspect of food quality does not appear to have been studied in the literature and therefore my subsequent analyses and comments are very tentative.

Let us first consider the dehydration of oil and fat containing foods such as o/w emulsions. Two broad cases are discussed.

A All flavour is retained

In this case the flavour release equations may be applied to follow the interphase transport of flavour as dehydration progresses. Table 4 presents

Table 4 Effect of the equilibrium partition coefficient, K_p, on interphase flavour transport during milk ($\phi_i = 0.035$) dehydration (Conditions: all flavour retained $DF_{em} = 0.111$; $C_{owi} = 100$ ppm; $V_{owi} = 10$ ml)

K_p	C_{wei} (ppm)	$\frac{C_{we}}{C_{wei}}$	$\% \frac{\bar{C}_{wei}}{\bar{C}_{ow}}$	$\% \frac{\Delta\bar{C}_w}{\bar{C}_{ow}}$
10^{-1}	103.0	12.2	99.4	3.80
10^0	100.0	9.00	96.5	28.1
10^1	76.0	3.09	73.3	55.4
10^2	22.4	1.25	21.6	19.5
10^3	2.78	1.03	2.68	2.46

the results of an application of the release equations to milk dehydration. These data show that major interphase flavour transport will occur if equilibrium conditions prevail especially in the range of K_p values 10^0–10^2. The latter covers the oil-water partition range of most flavour compounds.

It is possible, however, that equilibrium conditions are not attained especially as dehydration nears completion because the resistance to diffusion of flavour volatiles within the food matrix rises dramatically.[23] Thijssen proposed that volatiles are retained in food matrices by a "selective diffusion" entrapment mechanism whereas Flink and Karel[2] proposed a "micro-region entrapment" theory. In the latter case small pools of volatile have been entrapped and sealed in localized areas throughout the matrices of freeze-dried model systems. These micro-regions are very sensitive to moisture content. Above a certain critical level the seal dissolves and volatile loss commences. Thus during rehydration volatiles entrapped in micro-regions should be released rapidly whereas flavour dissolved in the oil or fat fraction may be released more slowly depending on the state of the oil or the fat and the type of food.

If the fat is in globular form in the range 0.5–5 μm and retains its globular form throughout dehydration, storage and rehydration (as in the case of milk), the flavour will also be released rapidly in accordance with the emulsion kinetic results previously presented. If, however, the fat is in bulk form then flavour release will be much slower and probably too slow to become available for perception in the mouth if consumption immediately follows rehydration.

Should, therefore, dehydration strategies attempt to entrap flavour within carbohydrate or other micro-regions rather than be dissolved or "fixed" in the fat fraction? This question is very difficult to answer at this time and requires experimental investigation. We can, however, state that the simple strategy of organizing processing conditions to maximize flavour retention need not necessarily result in a food with desirable flavour release characteristics.

B All flavour is not retained

If the percentage loss of each aroma compound due to dehydration is known then the distribution of the remaining flavours between oil and water can be readily evaluated by using C_{wei} values reduced by appropriate loss factors and repeating the calculations summarized in Table 4. Similar comments made in the previous case of full flavour retention apply here.

Finally it should be noted that the release equations may also be applied to rehydration as well as dehydration.

9 CONCLUSIONS

1 The flavour release equations revealed that: (a) flavour distribution or "fixation" in oils and fats increase as both ϕ_i and K_p increase; and (b) the potential extent of flavour release in the mouth increases as K_p, ϕ_i and/or emulsion dilution increase.

2 Stirred diffusion cell and emulsion studies revealed that the physicochemical properties of oils and fats (e.g. molecular composition, viscosity, solid fat index) and the presence of aqueous additives (e.g. salts, surfactants) can have a profound and complex effect on interphase flavour distribution, and on the potential extents and rates of flavour release in oil and fat containing food systems.

3 Application of the release model to *n*-alkanal threshold concentrations in oil and water[5] revealed that *n*-alkanal partition coefficients, and threshold concentrations in o/w emulsions (e.g. milk) could be satisfactorily predicted.

4 The release model was used to demonstrate why the "off-flavour" products of lipid oxidation are normally more easily perceived in o/w emulsions (e.g. milk) than in w/o emulsions (e.g. butter).

5 Many more instrumental and sensory data are required before the quantitative modelling methods presented here can be meaningfully applied to sensory perception of flavour in general.

6 More information is required on how processing operations such as dehydration affect flavour release characteristics of foods.

References

1. Feron, R. and Govignon, M. (1961). *Ann. Fals. Expert. Chim.* **54**, 308.
2. Flink, J. and Karel, M. (1970). *J. Agric. Fd. Chem.* **18**, 295.
3. Forss, D. A. (1972). *Progr. Chem. Fats Lipids*, **13**, 177.
4. Forss, D. A. (1981). Sensory characterisation in flavour research—recent advances, eds R. Teranishi, R. A. Fletcher and H. Sugisawa, Marcel Dekker, Basel, Switzerland.
5. Lea, C. H. and Swoboda, P. A. T. (1958). *Chem. Ind.* 1289.
6. Lillard, D. A., Montgomery, M. W. and Day, E. A. (1962). *J. Dairy Sci.* **45**, 660.
7. Lillford, P. J. (1984). Private communication. Unilever Research, Colworth, UK.
8. McDaniel, M. R., Sather, L. A. and Lindsay, R. C. (1969). *J. Fd Sci.* **34**, 251.
9. McNulty, P. B. (1972). Factors affecting flavour release and uptake in o/w emulsions. PhD Thesis. Massachusetts Institute of Technology, Cambridge.
10. McNulty, P. B. (1974). Flavour release from oils and fats. Mini-symposium Oils and Fats p. 29 IFST 10th Anniversary Symposium, London.
11. McNulty, P. B. (1975). *J. Pharm. Sci.* **64**, 1500.
12. McNulty, P. B. (1975). Quantitative analysis of flavour threshold data in various media using a flavour release model. *Procs. 4th Internat. Congr. Food Sci. & Technol.*, Madrid. Vol. II pp. 190–197.
13. McNulty, P. B. and Karel, M. (1973). *J. Fd. Technol.* **8**, 309.
14. McNulty, P. B. and Karel, M. (1973). *J. Fd. Technol.* **8**, 319.
15. McNulty, P. B. and Karel, M. (1973). *J. Fd. Technol.* **8**, 415.
16. McNulty, P. B. and Moskowitz, H. R. (1974). *J. Fd. Sci.* **39**, 55.
17. Nelson, P. E. and Hoff, H. E. (1968). *J. Fd. Sci.* **33**, 479.
18. Patton, S. (1962). Dairy Products in Lipids and Their Oxidation p. 190 H. M. Schultz, E. A. Day and R. D. Sinnhuber, eds. AVI Publ. Co. Inc., Westport, Connecticutt.
19. Patton, S. (1964). *J. Fd. Sci.* **29**, 679.
20. Salo, P. (1970). Determining the sensory levels of aroma compounds in alcoholic beverages. *J. Fd. Sci.* **35**, 95.
21. Siek, T. J., Albin, I. A., Sather, L. A. and Lindsay, R. C. (1969). *J. Fd. Sci.* **34**, 265.
22. Siek, T. J., Albin, I. A., Sather, L. A. and Lindsay, R. C. (1971). *J. Dairy Sci.* **54**, 1.
23. Thijssen, H. A. C. (1971). *J. Appl. Chem. Biotechnol.* **21**, 372.

13 Rheology Structure and Food Processing

P. RICHMOND and A. C. SMITH

AFRC Institute of Food Research, Norwich

1 INTRODUCTION

In recent years the food industry has undergone, and is continuing to undergo, significant change in response to external pressures from a number of sources. These pressures are responsible for the strong upturn in food research and especially the need to understand material properties, structures and rheology. Before discussing a particular application, it is helpful to understand in a little more detail the origin and nature of these changes which are occurring.

2 THE IMPACT OF HISTORICAL AND COMMERCIAL FACTORS UPON FOOD MANUFACTURE AND RESEARCH

The changes in question have occurred in large firms forming the apex of the industrial pyramid. There is also a thriving and numerically larger small food

FOOD STRUCTURE AND BEHAVIOUR
ISBN 0-12-104230-8

sector and concentration in product type and range amongst larger firms has increased their opportunities because it has left gaps in the market which large companies cannot fill economically. Medium sized companies occupy an uneasy position without the resources of the large firm, or the adaptability of the small firm, but in many cases they are in a state of transition and some are the large firms of the future. Of course, large firms have their origin in humble beginnings. For example, the Cadbury business started in 1824 with the opening of a grocer's shop in Birmingham. Their first advertisement drew attention to Cocoa Nips—a nutritious breakfast beverage. Soon the preparation of cocoa became the major part of the business and the product was sold on the strength of its reputation and the greatest problem became buying raw materials of high quality. The subsequent growth of firms like Cadbury in the 19th century was based on the appeal by the processor, directly to the customer, that his product or "brand" was of high quality. It was confidence in the manufacturer's reputation which led to increased sales, with wholesalers and retailers supplying goods demanded by the customers. The brand name, a guarantee of consistent quality, became the vehicle for growth. A food processor then had three concerns: (a) to gain control over the sources of supply to ensure consistency in raw materials; (b) efficient low cost production of finished goods; and (c) the selling of the product.

Expansion of sales overseas was a logical consequence of these concerns

UK Brand Leader

1933	Current position
Hovis, bread	No. 1
Stork, margarine	No. 1
Kellogg's cornflakes	No. 1
Cadbury's chocolate	No. 1
Rowntree's, pastilles	No. 1
Schweppes, mixers	No. 1
Brooke Bond, tea	No. 1
Colgate, toothpaste	No. 1
Johnson's, floor polish	No. 1
Kodak, film	No. 1
Ever Ready, batteries	No. 1
Gillette, razors	No. 1
Hoover, vacuum cleaners	No. 1

Source based on 1933 advertising expenditure and current market surveys.

Fig. 1

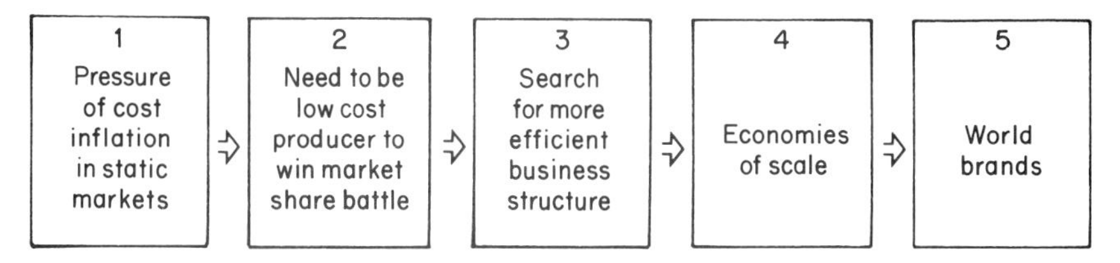

Fig. 2

but was also in response to the recognition that tastes in consumers in colonial countries such as Australia and New Zealand are similar to those of the UK and, there are groups of consumers whose tastes are similar to brand consumers in the UK, e.g. France, Germany and America. The international sales of Kelloggs cornflakes illustrates the world-wide appeal of certain branded food. The consequence of protectionism between the two World Wars was that food manufacturers established manufacturing operations overseas.

Since the Second World War, the food manufacturers, have become much more aware of their dependence on branded products and have intensively highlighted their quality aspects to obtain a premium price, countering increased costs and lower profit margins. Control over the supply of raw materials was reduced as governments at home and overseas increasingly legislated to influence the sale of agricultural products. This led to the need for alternative sources of materials consistent with "brand" quality. The increasing concentration in the retail trade with the development of supermarkets and abolition of retail price maintenance, has had a considerable impact on the concentration of food manufacturers. The evolution of retailers own "brands" from companies such as Sainsburys and Marks and Spencer, has prompted the manufacturers themselves to become suppliers of a wide range of products rather than one or two specialized products. All companies from brewers to confectioners began to have important customers in common. The larger companies began to move from batch production to continuous production and there was a convergence of technology across the food industry. Furthermore, other industries under similar pressures are broadening their base and beginning to use agricultural products and/or produce food materials. For example, both BP and ICI have considerable experience in making protein. The advances in biotechnology, process control, and flexible manufacturing systems will accelerate this trend to common technology.

The consequences for food research have been dramatic. Industries need to

Technical Challenge

- Novel products
- Flexible processes
- Alternative raw materials

Legislative Challenge

- Ingredients, additives,
  Processes not injurious
  To health and environment

Fig. 3 Challenges for food research

become low cost producers of high-quality products and this has meant the introduction of new technology and alternative raw materials linked with new product requirements. A further consequence has been the need to understand the material-process interaction in much more detail than previously, to ensure existing product quality and aid new product development. This requires input from engineering and physical sciences skills not traditionally associated with either the Food Industry or Food Research but well established in the oil and chemical industries. The development of electronic information systems will also provide distributors and manufacturers with much economic information and allow detailed economic and market experiment and research.

In thc following sections we outline a general approach to the study of processing in the context of a particular food product—namely maize or corn flakes. Section three extends these considerations and presents as a wider canvas some results of work on extrusion of maize, covering processing, microstructure and mechanical properties. Finally, we discuss some aspects of the creation of filled anisotropic composites.

3 PROCESSING, PROPERTIES AND STRUCTURE

A General considerations

In a previous paper (Atkins p. 149) the relationship between structure and mechanical properties of natural materials has been considered. Generally the emphasis is on making materials harder and/or tougher. Such properties are one set that are important in food products. The difficulty is in specifying the user or (sensory) properties and their link to raw materials (cf Fig. 4).

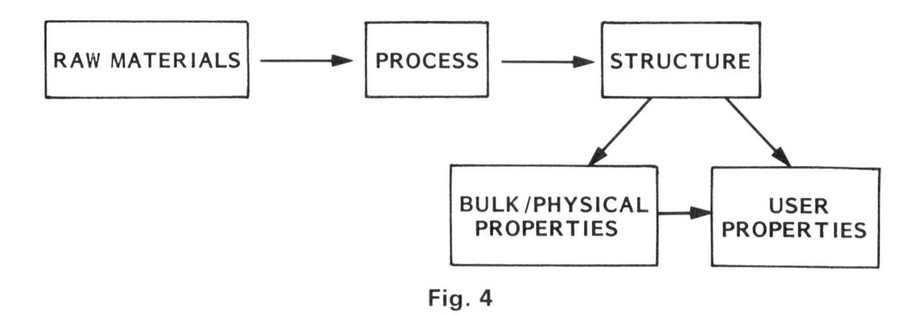

Fig. 4

In a number of these systems, these properties can be traced to characteristic structural elements, their interaction with each other, and modification via environmental changes, e.g. moisture/shear, etc. In food manufacturing, the aim is to mimic existing products/structures *via* new assembly routes or with alternative materials; or to produce new product structures with desirable properties.

The key to a good product is the specification of the raw materials and the appropriate physical and/or chemical steps that comprise the process and yield the desired structure. These can involve a variety of transfer processes such as heat, mass and momentum transfer, as well as chemical or biological reactions and structural or phase changes.

B Cereal processing and structure

To illustrate these somewhat abstract concepts, consider corn flake manufacture. The classical process where whole corn or maize is transformed into a flake is a batch process. This process involves mass and heat transfer in steam cooking, cooling and drying prior to momentum transfer during roller milling and, finally, intensive heat transfer during toasting prior to packing (Fig. 6).

Some companies have developed their brand around products produced in this way. However, is it possible to produce such flakes using alternative (cheaper/more handleable) raw materials and/or other technology that offers more opportunities for control than the somewhat space intensive batch process?

The breakdown of the batch process in terms of transfer processes suggests that the equipment is not unique in its processing ability. In recent years the food industry has been examining the use of extruders for transforming maize grits into products such as flakes. One such process line is shown in Fig. 8.

This uses an extruder to continuously form a pellet of cooked, gelatinized

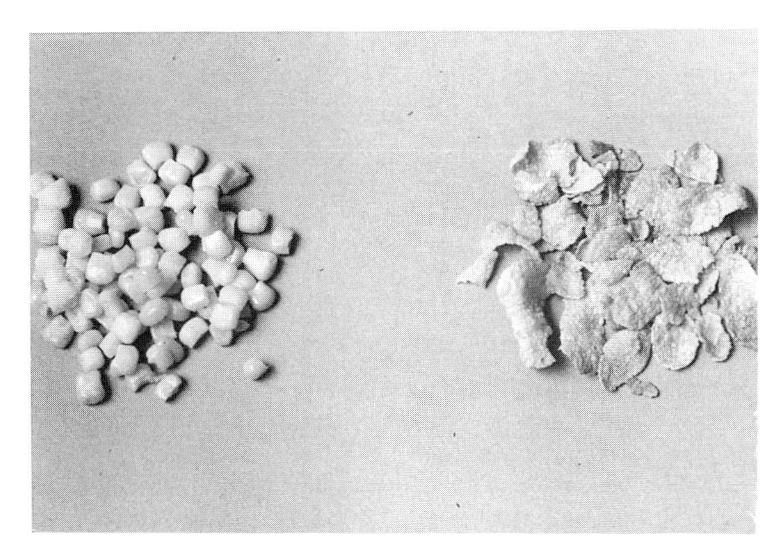

Fig. 5 Corn or maize and corn flakes

material which can then be handled in a similar manner to the previous process. Another more simplified process is shown in Fig. 9.

This combines all the transfer processes into one piece of equipment, the extruder. The maize grits are gelatinized and formed into a dough in the extruder as before, but now the structuring process occurs as the pressure

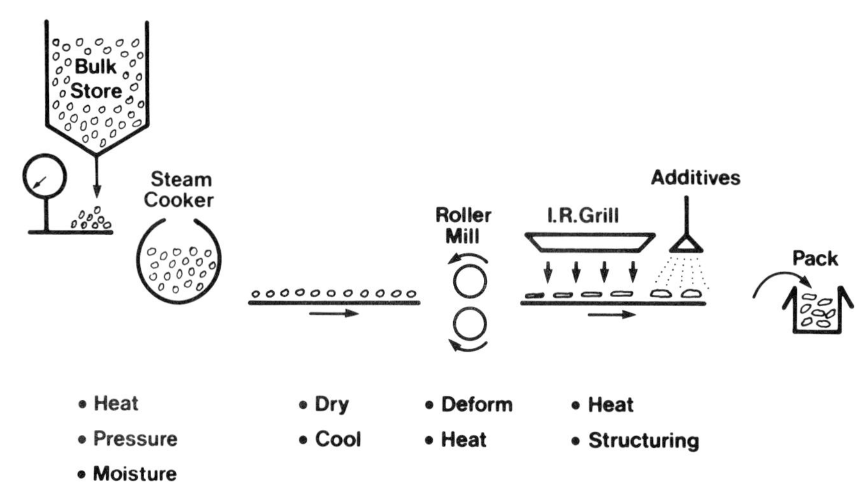

Fig. 6 Batch corn flake manufacture

Fig. 7 Corn or maize grits and corn flakes

drops at the exit die and included moisture flashes off, expanding the structure. This feature has, in fact, led to a variety of novel products apart from the familiar corn flake.

Using both of these processes, it is possible to manufacture products that

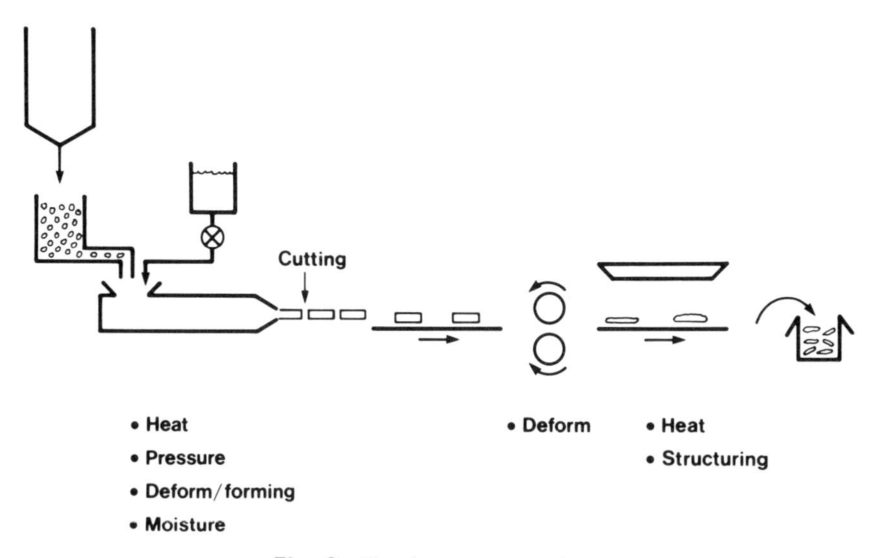

Fig. 8 Continuous processing

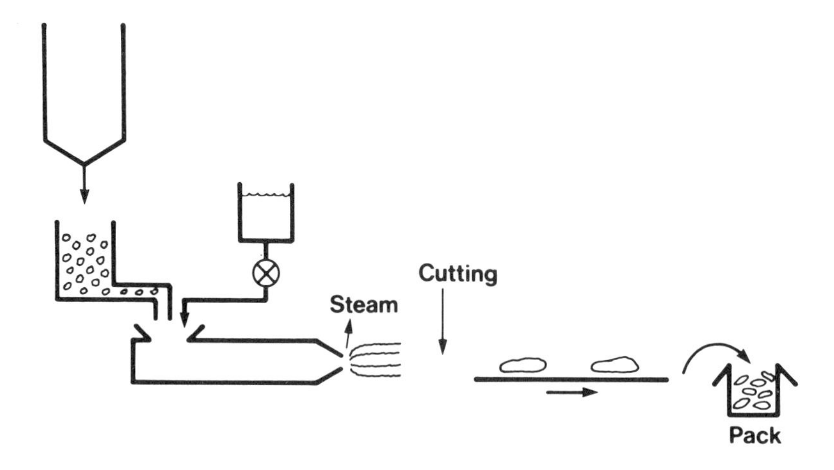

- **Heat**
- **Pressure**
- **Deformation**
- **Evaporation**

Fig. 9 Single stage continuous processing

are similar in appearance and flavour. Yet products from the processes can be perceived to be different by the consumer. This difference is quantified in textural terms, and can be related to the interaction between the product and moisture under simultaneous shear. Given that the basic molecular composition of raw and processed material in terms of protein, starch, lipid, etc., is, unchanged, and the processed macrostructures on a visual scale of scrutiny (10^{-3} m) are similar, a logical conclusion is that the microstructure is being modified.

With our cereal or corn flake, a variety of microstructures which influence the physical properties can be produced depending on process history. Figs 10 and 11 illustrate this point with differently processed maize examples.

The controlled manufacture of our corn flake can now be resolved into a number of different aspects. First, we require a detailed study of the material microstructure to identify the nature and interaction of the intermediate structure. This must be linked to a study of the influence of mechanical deformation, temperature changes, etc. Finally, the link to bulk rheology and non-rheometric flows that occur in particular pieces of equipment such as extruders must be made for rational process design. This approach is illustrated in Fig. 12 where a subjective view of the relationships between approach and skills is presented.

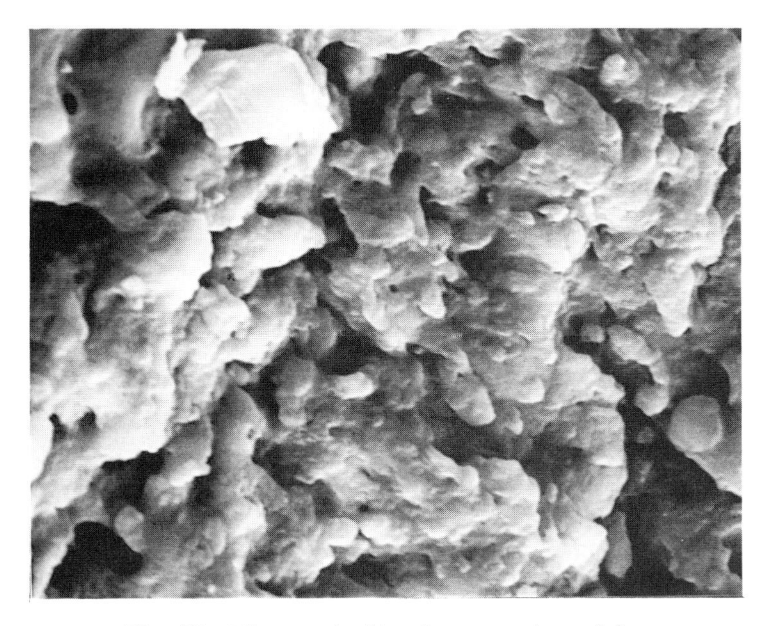

Fig. 10 Micrograph of batch processed cornflake

C Cereal processing and raw material properties

We have noted above that extrusion cooking is a rapidly expanding technique for continuously processing food materials. It may be used as a part or a whole process to create new physical and chemical structures from a number of raw or partially cooked ingredients. The cooking extruder subjects the ingoing material to a complex history comprising high temperatures, pressures, strains and strain rates in a short time. This complexity has led to a situation where an understanding of the extrusion cooking process lags well behind its widespread use. One of the first uses of extrusion cooking was concerned with cereals, the majority of these studies involved single screw extruders.[1-5] Most work has involved the characterization of the extruded product in terms of colour, expansion, bulk density, water solubility and absorption and impact properties. The aqueous viscosity of extruded cereal dispersions has principally been examined using amylograph techniques.[4-6] For example, the viscosity of maize extrudate pastes has been studied and a marked influence of screw configuration, moisture and barrel temperature observed.[6] The variation of aqueous paste viscosities with heating sequences has led to a suggestion that the degree of cooking of extruded cereal products may be deduced.[7] The viscosity response of gelatinized and dextrinized maize samples has also been studied and used in a prediction of extruded maize

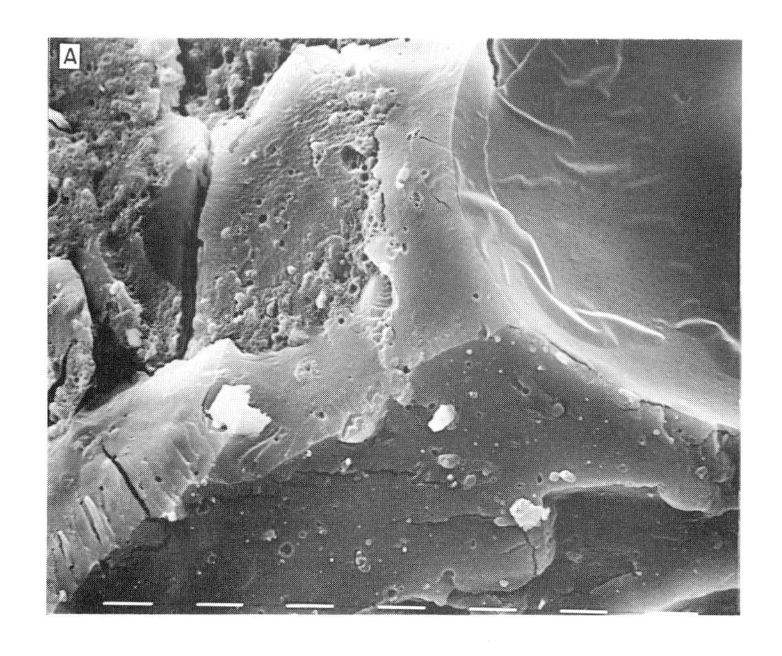

Fig. 11*A*, 11*B* Micrographs of flakes extruded under different conditions

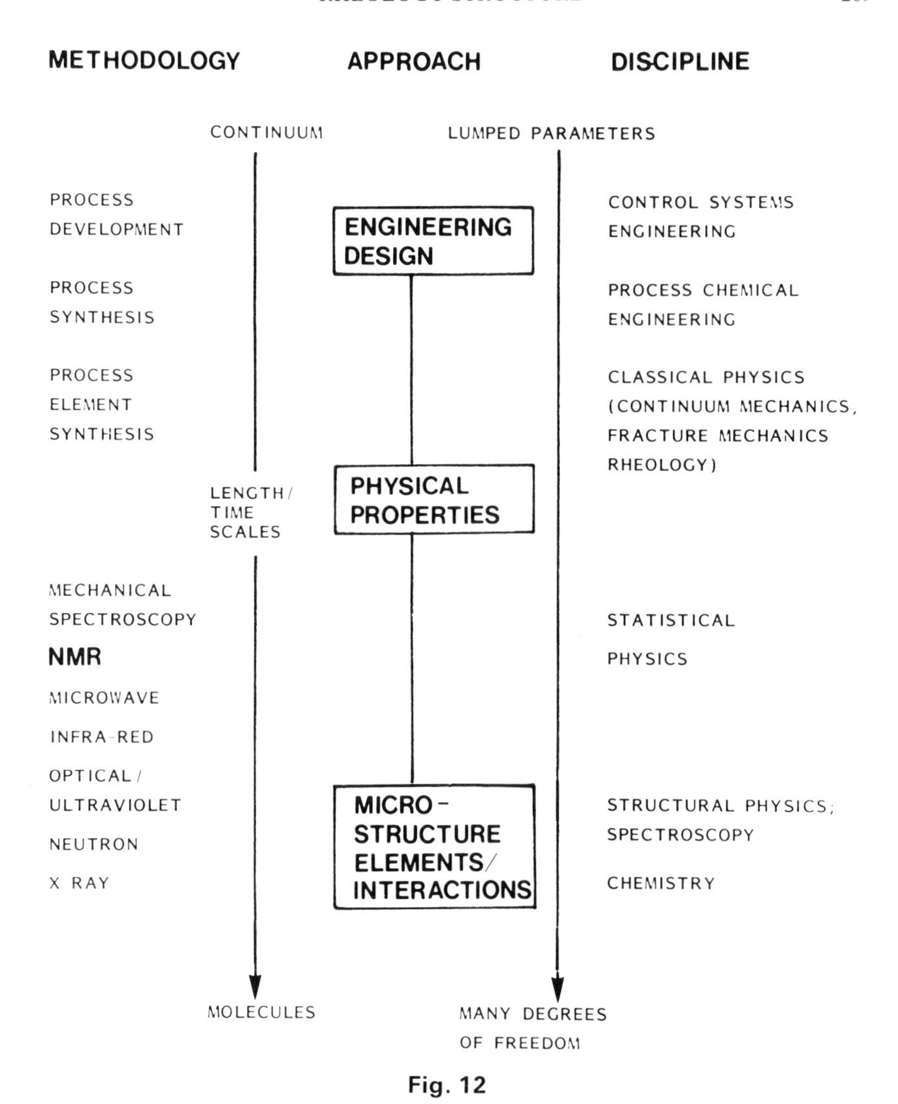

Fig. 12

viscosity.[8] This approach describes the principal outcome of extrusion cooking in terms of starch gelatinization and depolymerization of the maize macromolecular structure.

The starch component of cereals has received more attention using twin-screw extrusion.[9–11] Extruded maize starch has been treated using an amylograph heating sequence prior to determination of its viscosity using Couette flow rheometry.[12] More commonly the intrinsic viscosity of

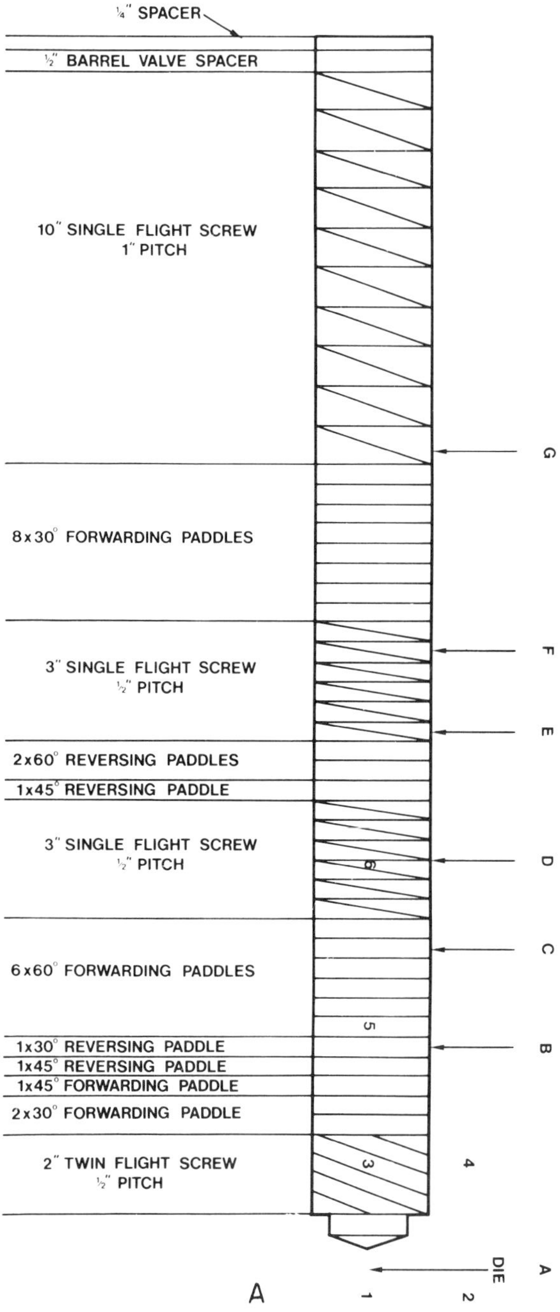

¼" SPACER

½" BARREL VALVE SPACER

10" SINGLE FLIGHT SCREW
1" PITCH

G

8 x 30° FORWARDING PADDLES

3" SINGLE FLIGHT SCREW
½" PITCH

F

E

2 x 60° REVERSING PADDLES

1 x 45° REVERSING PADDLE

3" SINGLE FLIGHT SCREW
½" PITCH

6

D

6 x 60° FORWARDING PADDLES

C

5

1 x 30° REVERSING PADDLE

B

1 x 45° REVERSING PADDLE

1 x 45° FORWARDING PADDLE

2 x 30° FORWARDING PADDLE

2" TWIN FLIGHT SCREW
½" PITCH

3

4

DIE

A

A

1

2

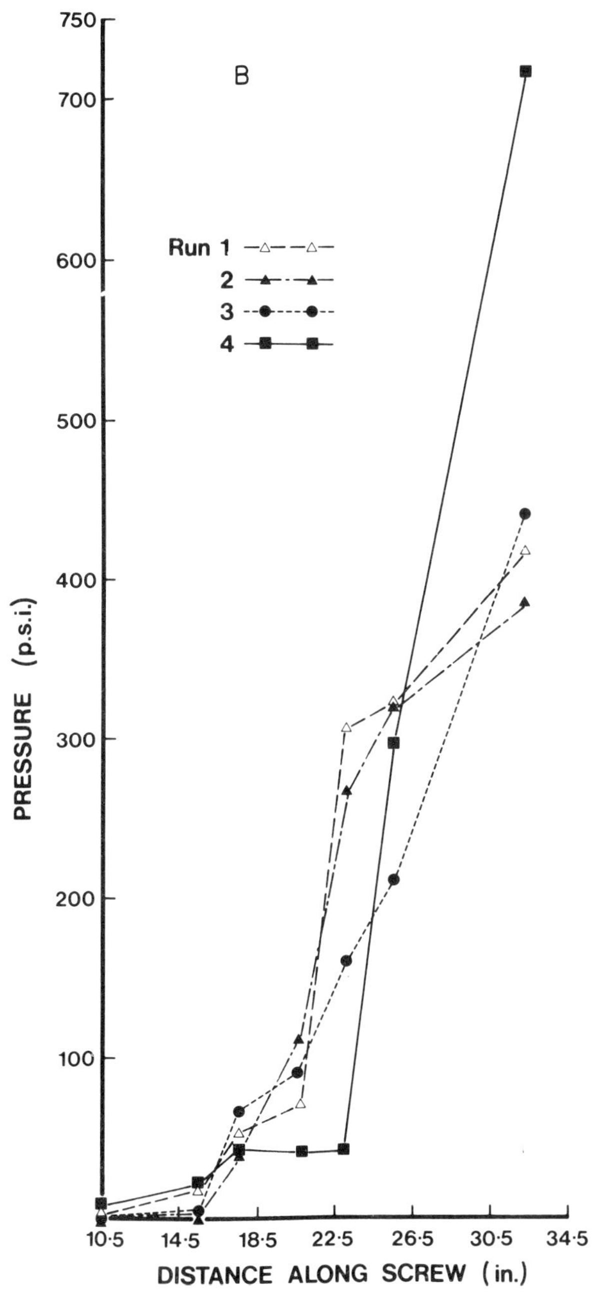

Fig. 13*A*, 13*B* Typical twin-screw configuration and associated pressure profiles

extruded starches has been studied. A comparison of dispersion viscosity and intrinsic viscosity of extruded starches has been carried out using a twin-screw extruder.[13] The intrinsic viscosity of partially extruded starches has been examined using samples removed from the barrel following a dead-stop procedure.[9-11]

Work in the Process Physics Division at the Food Research Institute, Norwich has considered the viscosity of maize dispersions for extruded and partially extruded material. The aqueous dispersion viscosity before and after treatment to 80 °C for samples removed from a twin-screw extruder following a dead stop procedure has been studied.[14] This work also indicated that the extrudate dispersion rheology was a sensitive indicator of the extrusion conditions in agreement with the degree of cooking experiments.[7]

Typically in these experiments commercial maize grits of 600 μm nominal size were extruded using a Baker Perkins MPF 50 intermeshing, co-rotating twin-screw extruder. The extruder barrel could be opened and the screws built up from various conveying and paddle elements that are used to control residence time and the degree of shear applied to the material. Five set temperature values can then be used with each screw configuration with appropriate maize and water feed rates for stable operation of the extruder. The extruder was instrumented to measure the torque, screw speed and feed rates which were logged using a microcomputer system. Typical pressures and temperatures along the extruder barrel were also continuously monitored at the points A to G shown in Figs. 13*A*.

The equilibrium pressure profiles show the barrel pressure does not begin to rise until the final third of the barrel length, corresponding to sample point 6 (Fig. 13*B*). At this stage the maize has formed a compacted solid in which the grits are surrounded by a maize film adjacent to the barrel wall. The film grows at the expense of the grits until a melt or dough is formed. This process is accompanied by colour changes that have been described elsewhere.[15] A uniform dough does not form until sample point 4 by which time the pressure has increased dramatically. At sample point 5 some melting of the maize within the C-shaped chambers of the extruder has occurred giving rise to inhomogeneous samples. The pressure in a twin extruder has been predicted to develop when the chambers become full. Viscous material in the extruder would result in a linear pressure increase towards the die.[16] The present results show that the pressure begins to increase when the solid material is compacted and that the pressure rises towards the die independently of the phase changes that occur. Again more detailed analysis of the pressure profiles for these and other experiments is given elsewhere.[17]

Under steady conditions of pressure, torque and temperature samples of the extrudate were collected. The extruder was stopped by cutting the feed supply, the electrical heating and the screw rotation. The barrel was then cooled by the refrigeration system to ambient temperature and dismantled (Fig. 14). Samples were removed from the extruder barrel at the locations 1 to 6 shown in Fig. 13.

The transformation of the maize microstructure as a function of progress along the extruder barrel is shown in the SEM sequence of Fig. 15. The primary transformation during extrusion is shown in the starch component. The raw maize contains starch granules of 10–15 μm in diameter. With

Fig. 14 View of open extruder barrel. The transition from compacted powder to melt is marked by the darker colour of the material in the screw

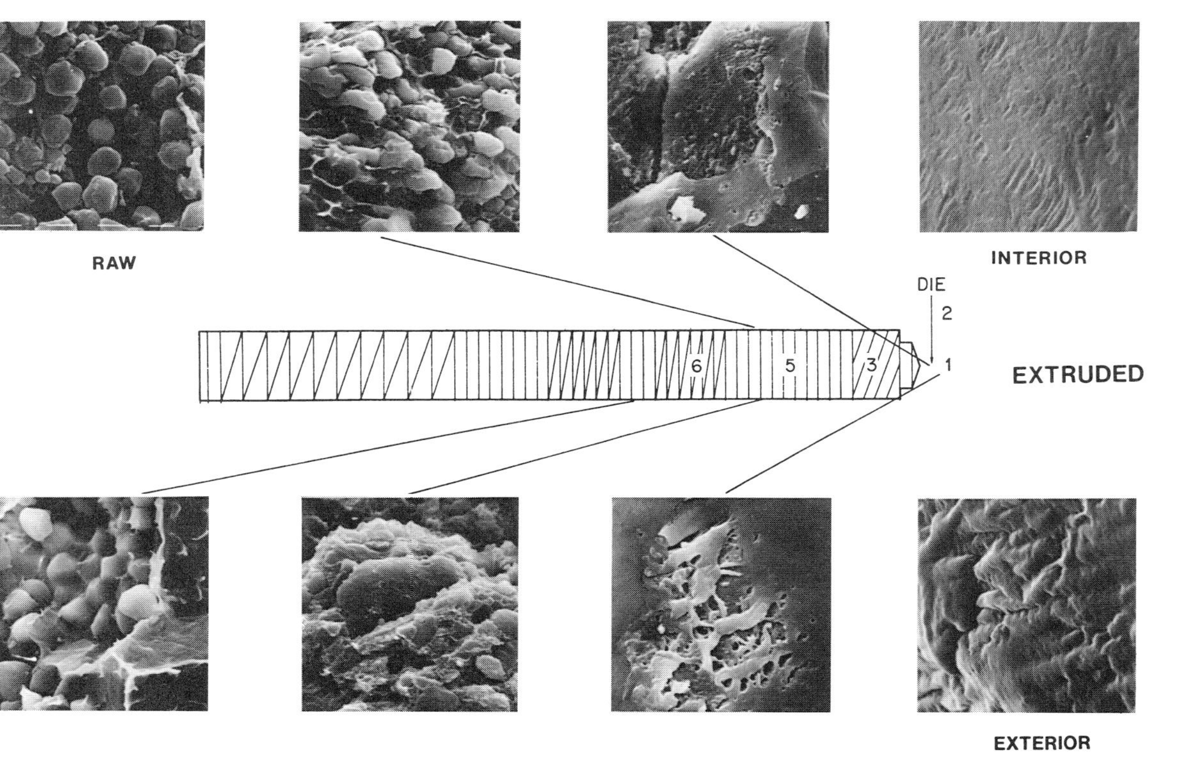

Fig. 15 Micrographs of the structure change associated with the conditions shown in Fig. 13

progress along the extruder the granules melt together and deform and some granules are plucked from the protein matrix. This corresponds to a temperature rise from 95 °C–140 °C and a pressure rise from 40–150 psi for this experiment. Further starch granule deformation occurs until a new texture is formed at the die at a temperature in excess of 160 °C and a pressure of 440 psi. Linked to these structural studies, it is of interest to carry out parallel rheological studies of dilute dispersions of milled material. These were prepared in a centrifuge mill to within 180–250 μ.

The milled material was added to distilled water to give a 9.1% dry weight dispersion which was stirred for 30 min. The dispersion was then sheared in a Couette rheometer (Contraves Rheomat 115) according to the following shear history: (a) the rate was increased to 148 s^{-1} in 431 seconds; (b) the rate was maintained constant at 148 s^{-1} for 120 sec; (c) the rate was decreased linearly to zero in 431 sec. The dispersion was then heated and stirred for 15 min at 40 °C followed by repetition of the above shear procedure at ambient temperature. This was repeated for heat treatment at 60, 70 and 80 °C.

The dispersion viscosity at the maximum shear rate of testing (148 s^{-1}) is plotted as a function of maximum heat treatment or cooking sequence for the different extracted samples in Fig. 16.

The viscosity of the raw maize dispersion (Fig. 16(D)) increases by at least an order of magnitude on heat treatment to 80 °C. The principal cause is the gelatinization of the starch component of maize although increased solubility

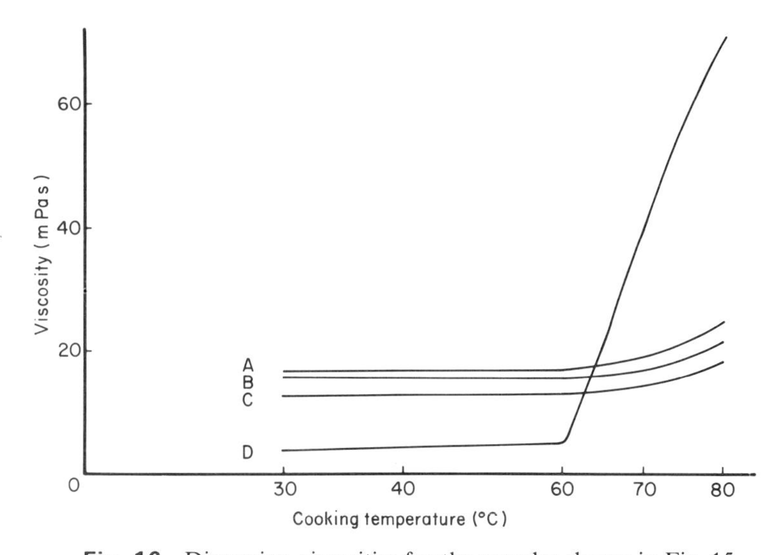

Fig. 16 Dispersion viscosities for the samples shown in Fig. 15

would be contributory. The extrudate dispersion viscosity (Fig. 16(A-C)) shows only a slight increase with heat treatment which may be attributed to an increase in solubility. There may be some gelatinization effects in the extruded maize owing to the much higher water availability of the dispersions although discrete granule swelling is no longer possible. Furthermore, the interaction of the maize constituents such as starch, proteins and lipids with each other and with the greater availability of water in the dispersions is unknown and may further complicate their rheology.

Other work[18] shows that the dispersion viscosity is relatively insensitive to the extrusion shear rate, both before and after heat treatment in an excess of water. This implies that the solubility and swelling levels in the extrudate are little changed.

Some thermodynamic studies have been made of structural changes in the starch granules during heating as a function of moisture and pressure.[19,20] However there is clearly much more to be done before a complete picture can be put together.

Turning to the user properties of these products it is clear that these are related in a complex way to the composition, structure and physical properties. However the bulk density, porosity and mechanial properties form a subset of important characteristics that determine the so-called "crunch" factor of expanded products and we have begun to examine how these relate to the overall structure of expanded extrudate. From studies using a slit die it appears that the bulk density is proportional to the dough viscosity in the die and is independent of shear rate and temperature.[18] An expanded product structure can be created as we noted above through the "flashing" of steam from superheated water as the pressure falls to atmospheric at the die exit. The bulk density would be expected to depend on the water content and temperature of the dough in the die. The available data suggest that the viscosity determines the expanded volume. This escape of steam from the extrudate causes an expansion of the product which is governed by its mechanical properties. A viscous dough will resist deformation by the gas with the result that a low volume, high bulk density product is formed. The dough viscosity takes into account the effects of reactions, moisture and shear rate in the dough which are not described by temperature alone.

Extreme examples of the resulting structure fractured for microscopy are shown in Fig. 17. These low power SEMs which are both the same magnification clearly illustrate the contrasting structures that can be produced by appropriate processing. We have made some preliminary mechanical tests using an impact device based on an instrumental pendulum. As one would expect, these two structures in question behave quite differently. Both appear to undergo brittle failure and therefore it may be possible to apply the simple ideas of Griffiths[21] to interpret their failure

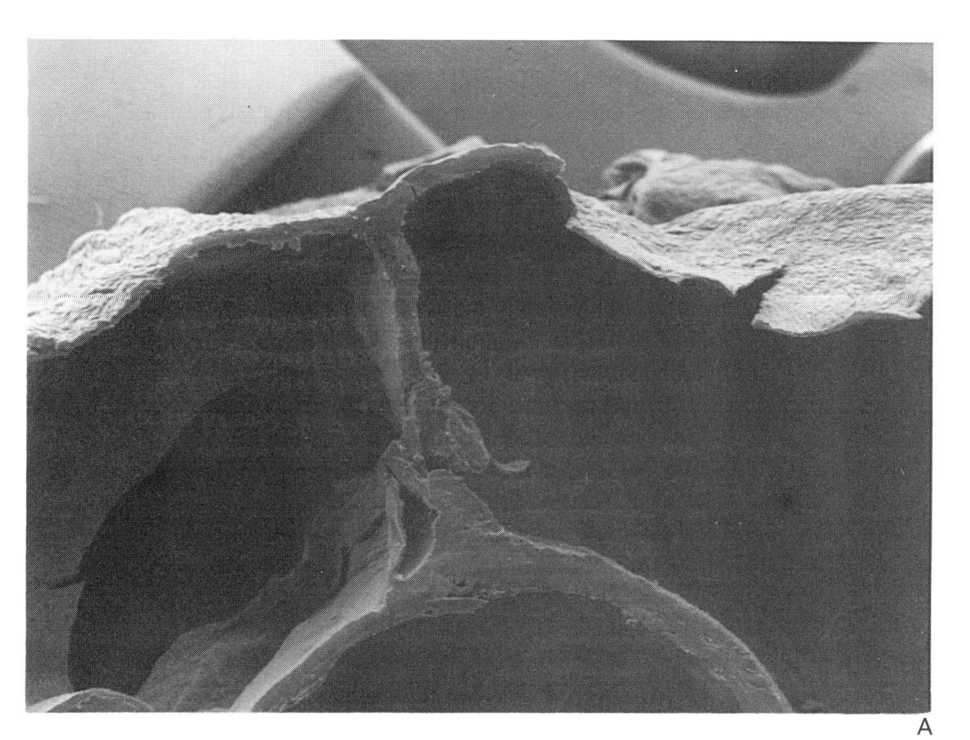

A

B

Fig. 17*A*, 17*B* Low power micrographs of identical material produced under different processing conditions via extrusion

properties. We shall report these results in more detail elsewhere in conjunction with parallel work on consumer properties.

4 THE DEVELOPMENT OF NEW TEXTURES

A The example of filled composites

So far we have discussed the processing and properties of "simple" cereal materials. However, there are two major reasons for looking in more detail at the influence of other components. The first is related to the need to improve production or output efficiency; the second follows from the requirements of marketing for new concepts that will boost sales.

To improve performance, use has been made of additives such as glycerides—particularly monoglycerides—to improve viscosity and frictional characteristics of the raw and processed materials. One school of thought is that the amylose and monoglyceride can form a more complex geometric structure. This may then modify the opportunities for recrystallization of the starch and retrogradation or staling in, for example, bread. With regard to processing, it may simply be mixing/lubrication that causes the modified performance. However, a detailed explanation for the influence on the viscosity and frictional parameters remains to be found.

The requirements of marketing can be many and varied. An ability to produce products filled with secondary material, e.g. raisins, nuts etc., can create interesting textures. The ability to sell "nutrition" in a quantitative manner could give a company a significant advantage. The strong interest in dietary fibre stems from this. Dietary fibre is, essentially, cell wall material that is not digested and can be readily obtained as flakes. Preliminary tests with these particles suggest they can retain their integrity during extrusion and it is of interest to examine how they interact with the structure and properties of the extruded materials we have discussed above.

If such particles are introduced at a low level into the process feed material, then to a first approximation they will, on entering the extruder, follow the flow or stream lines and, at a steady state, will emerge more or less evenly distributed. However, for such anisotropic particles, orientation effects occur and these can modify the final material properties. It is instructive to examine this in more detail and compare, in particular, the different effects of shear and extensional flow on orientation.

Consider, for simplicity, needle-like particles which can be characterized by a director n embedded in a velocity field, $u(r)$. For particles embedded in a highly viscous medium it is reasonable to neglect inertia. Ignoring other interactions on the embedded particles, their orientation as a function of time can then be determined from the equation

$$\frac{\mathrm{d}n}{\mathrm{d}t} = [u(r+n) - u(r)][l - nn]$$

subject to an appropriate initial condition. This automatically satisfies the

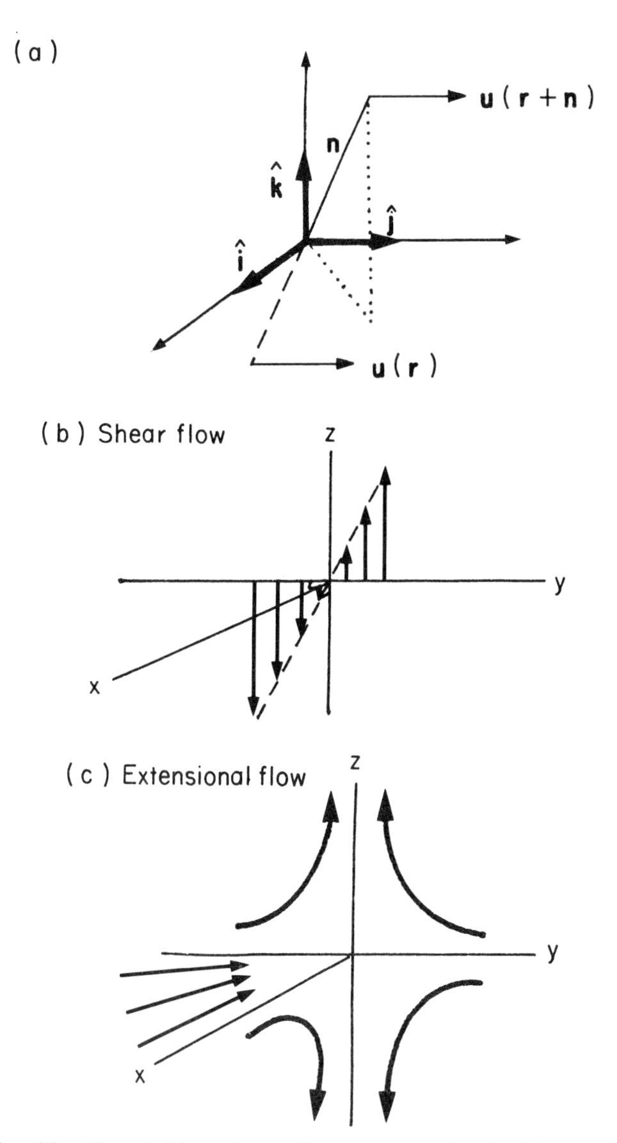

Fig. 18 Flow fields and coordinate systems for the "director"

constraint $d(n^2)/dt = 0$. For "point" directors we can expand the first term in brackets to obtain:

$$\frac{dn}{dt} = n.e - n(n.e.n) \qquad \text{where } e = \nabla u \tag{1}$$

1 SHEAR FLOW

With the usual vector notation shear flow in the direction of the z axis may be written

$$v = \gamma y k \tag{2}$$

The velocity gradient is then

$$e = \gamma j k \tag{3}$$

where γ is the shear rate.

The director may be expressed in spherical polars:

$$n = \sin\theta \cos\phi\, i + \sin\theta \sin\phi\, j + \cos\theta\, k \tag{4}$$

Using equations 1–3 it can be shown that

$$\dot{\theta} = -\gamma \sin^2\theta \sin\phi \tag{5}$$

$$\dot{\phi} = 0 \tag{6}$$

It follows immediately from equations 5 and 6 that the rate at which particles orient in a shear flow depends on the azimuthal angle ϕ but this angle itself does not change.

Integrating equation 4 gives

$$\theta = \cot^{-1}[\gamma t \sin\phi + \cot\theta_0] \tag{7}$$

where θ_0 is the initial orientation. Of interest is the asymptotic behaviour at large times. It is straightforward to see that:

$$\theta \sim 1/t \tag{8}$$

unless $\phi = 0$ or π in which case the initial orientation will be unaltered. However it is easy to see from equation 7 that the direction $\theta = 0$ is an unstable point and if the director is displaced to a negative value it will rotate through π. Such displacements could arise in colloidal systems from diffusion; in true microscopic systems interactions or collisions between directors could generate such displacements. However for an assembly of rods there will be a net orientation and for sufficiently strong interactions between rods the orientation under flow can be expected to be enhanced. Such orientation effects have recently been observed using light scattering by Penfold.[22]

2 EXTENSIONAL FLOW

Using the same notation, an extensional flow cylindrically symmetric about the z axis may be written

$$v = -\varepsilon xi - \varepsilon yj + \varepsilon zk \tag{9}$$

where ε is the extension rate. The velocity gradient field is then:

$$e = -\varepsilon ii - \varepsilon jj + \varepsilon kk \tag{10}$$

The equation of motion 1 now yields

$$\dot{\theta} = -\varepsilon \sin 2\theta \tag{11}$$

$$\dot{\phi} = 0 \tag{12}$$

As with shear flow ϕ does not change but now note that θ is independent of ϕ as one would expect. For $\theta = \pi/2$, equation 11 yields $\dot{\theta} = 0$, so particles on the xy plane perpendicular to the stretching motion would remain in position. However, as with shear flow one would expect either fluctuations from thermal effects or collisions to ensure total disorientation.

Integrating equation 11 now yields:

$$\theta = \tan^{-1}[\tan \theta_0 e^{-2\varepsilon t}] \tag{13}$$

For large t ($\varepsilon t \gg 1$, $\theta_0 \neq \pi/2$) this yields the asymptotic relation:

$$\theta \sim e^{-2\varepsilon t} \tag{14}$$

It is immediately evident that extensional flow is a much stronger flow and creates total orientation much more efficiently than shear. This suggests, for example, that suitably designed converging dies can be a more effective means of orientation than slit or cylindrical dies in extrusion.

In general, an assembly of particles will have an initial distribution $P\ (nt_0)$. For example, for a well mixed system $P\ (nt_0) = 1/4\pi$ (note P is normalized $(nt)\ dn = 1$). The equation of motion for P follows from the Liouville equation:

$$\frac{\partial P}{\partial t} + \nabla . J = 0 \tag{15}$$

where the flux

$$J = nP + J_R \tag{16}$$

In the absence of stochastic effects, $J_R = 0$. However, as we have already noted, for colloidal particles collisions will occur and, generally, there will be fluctuations in the flow. This will lead to the form $J_R \sim D\nabla_n P$ where the coefficient D depends on the nature of the fluctuations. This type of equation has been studied in detail for some colloidal systems, e.g. rods and macromolecules[23] but little work has been done on completely classical systems where collisions and flow dominate the interactions. In this case, one expects the equation of motion (1) to be modified and include terms of the form $-\lambda \sum n_i . n_j$ where λ is a measure of the excluded volume interaction and the subscripts refer to the different directors or rods. These may be approximated by a mean field type approximation to yield $-\lambda n . \langle n \rangle$ where the angular brackets denote an ensemble average. The problem is now similar to an orthodox many body problem but instead of temperature, the flow field itself must give the appropriate independent parameter that determines the average statistical configuration. The statistical mechanical analogy is now with information theory.[24]

Acknowledgement

The authors would like to thank R. J. Turner for obtaining the scanning electron micrographs.

References

1. Anderson, R. A., Conway, H. F., Pfeifer, V. F. and Griffin, E. L. (1969). *Cereal Sci. Today* **14**, 4–11.
2. Harper, J. M., Rhodes, T. P. and Wanninger, L. A. (1971). *Chem. Eng. Progr. Symp.* **67**, 40–43.
3. Van Zulilchem, D. J., Buisman, G. and Stolp, W. (1974). Shear Behaviour of Extruded Maize, IUFOST Conference, Madrid (1974).
4. Mercier, C. and Feillet, P. (1975). *Cereal Chem.* **52**, 283–297.
5. Van Zuilichem, D. J., Bruin, S., Janssen, L. P. B. M. and Stolp, W. (1980). Single Screw Extrusion of Starch- and Protein-Rich Materials in Food Process Engineering Vol. 1 ed. Linko, P., Malkki, Y., Olkku, J. and Larinkari, J. Applied Science, London.
6. Seiler, K., Weipert, D. and Seibel, W. (1980). Viscosity Behaviour of Ground Extrusion Products in Relation to Different Parameters in Food Process Engineering Vol. 1 ed. Linko, P., Malkki, Y., Olkku, J. and Larinkari, J. Applied Science, London.
7. Paton, D. and Spratt, W. A. (1981). *Cereal Chem.* **58**, 216–220.
8. Gomez, M. H. and Aguilera, J. M. (1983). *J. Fd Sci.* **48**, 378–381.
9. Mercier, C., Charbonniere, R., Gallant, D. and Guilbot, A. (1979). Structural Modification of Various Starches by Extrusion Cooking with a Twin-Screw French Extruder in Polysaccharides in Food ed. Blanshard, J. M. V. and Mitchell, J. R., Butterworth.
10. Colonna, P. and Mercier, C. (1983). *Carbohydrate Polymers* **3**, 87–108.
11. Colonna, P., Melcion, J.-P., Vergnes, B. and Mercier, C. (1983). *J. Cereal Sci.* **1**, 115–125.
12. Launay, B. and Lisch, J. M. (1983). *J. Fd Eng.* **2**, 259–280.

13. Launay, B. and Kone, T. (1984). In Thermal Processing and Quality of Foods, ed. Zeuthen, P. *et al.*, Applied Science, London, pp. 54–61.
14. Fletcher, S. I., McMaster, T. J., Richmond, P. and Smith, A. C. (1984). In Thermal Processing and Quality of Foods, ed. Zeuthen, P. *et al.*, Applied Science, London, pp 223–234.
15. Fletcher, S. I., McMaster, T. J., Richmond, P. and Smith, A. C. (1984). In Physics and Extrusion of Soft Solid Foodstuffs, ed. McKenna, B., Applied Science, London.
16. Janssen, L. P. B. M. (1978). Twin Screw Extrusion, Elsevier, Amsterdam.
17. Richmond, P. and Smith, A. C. (1985). *Brit. Poly. J.* **17**, 246–250.
18. Fletcher, S. I., McMaster, T. J., Richmond, P. and Smith, A. C. (1985). *Chemical Engineering Communications.* **32**, 239–262.
19. Takahashi, K., Shirai, K. and Wada, K. (1982). *Agric. Biol. Chem.* **46**, 2505–2511.
20. Muhr, A. H., Wetton, R. E. and Blanshard, J. M. V. (1982). *Carb. Poly.* **2**, 91–102.
21. For an introduction to Griffiths ideas, see, for example, the excellent text by Gordon, J. E. The New Science of Strong Materials, Pelican, 1977.
22. J. Penfold, (to be published).
23. Doi, M. and Edwards, S. F. (1978). *J. Chem. Soc. Far. Trans. II* **74**, 560; *Ibid* **74**, 918.
24. See for example P. Richmond, (1978). *Math. Scientist* **3**, 63–82.

Index